# THE OXFORD HISTORY OF ENGLISH LITERATURE

*General Editors*

JOHN BUXTON, NORMAN DAVIS, BONAMY DOBRÉE, *and* F. P. WILSON

V

# THE OXFORD HISTORY OF ENGLISH LITERATURE

Certain volumes originally appeared under different titles (see title-page versos). Their original volume-numbers are given below.

I. Middle English Literature 1100–1400
J. A. W. Bennett; edited and completed by Douglas Gray
*(originally numbered I part 2)*

II. Chaucer and Fifteenth-Century Verse and Prose
H. S. Bennett *(originally numbered II part 1)*

III. Malory and Fifteenth-Century Drama, Lyrics, and Ballads
E. K. Chambers *(originally numbered II part 2)*

IV. Poetry and Prose in the Sixteenth Century
C. S. Lewis *(originally numbered III)*

V. English Drama 1485–1585
F. P. Wilson; edited with a bibliography by G. K. Hunter
*(originally numbered IV part 1)*

VI. English Drama 1586–1642: Shakespeare and his Age
*(forthcoming; originally numbered IV part 2)*

VII. The Early Seventeenth Century 1600–1660: Jonson, Donne, and Milton
Douglas Bush *(originally numbered V)*

VIII. Restoration Literature 1660–1700: Dryden, Bunyan, and Pepys
James Sutherland *(originally numbered VI)*

IX. The Early Eighteenth Century 1700–1740: Swift, Defoe, and Pope
Bonamy Dobrée *(originally numbered VII)*

X. The Age of Johnson 1740–1789
John Butt; edited and completed by Geoffrey Carnall
*(originally numbered VIII)*

XI. The Rise of the Romantics 1789–1815: Wordsworth, Coleridge, and Jane Austen
W. L. Renwick *(originally numbered IX)*

XII. English Literature 1815–1832: Scott, Byron, and Keats
Ian Jack *(originally numbered X)*

XIII. The Victorian Novel
*(forthcoming; originally numbered XI part 1)*

XIV. Victorian Poetry, Drama, and Miscellaneous Prose 1832–1890
Paul Turner *(originally numbered XI part 2)*

XV. Writers of the Early Twentieth Century: Hardy to Lawrence
J. I. M. Stewart *(originally numbered XII)*

# ENGLISH DRAMA
# 1485–1585

F. P. WILSON

*edited with a bibliography by*

G. K. HUNTER

CLARENDON PRESS · OXFORD

*Oxford University Press, Walton Street, Oxford* OX2 6DP
*Oxford New York Toronto*
*Delhi Bombay Calcutta Madras Karachi*
*Petaling Jaya Singapore Hong Kong Tokyo*
*Nairobi Dar es Salaam Cape Town*
*Melbourne Auckland*
*and associated companies in*
*Berlin Ibadan*

*Oxford is a trade mark of Oxford University Press*

*Published in the United States*
*by Oxford University Press, New York*



*First published 1968*
*Reprinted 1979, 1990*

*Originally published as volume IV part 1 (ISBN 0-19-812209-8)*



*British Library Cataloguing in Publication Data*
*data available*

*Library of Congress Cataloging in Publication Data*
*data available*
*ISBN 0-19-812232-2*

*Printed in Great Britain by*
*Courier International*
*Tiptree, Essex*

# PUBLISHER'S NOTE

## F. P. WILSON 1889–1963

WHEN the plan of an Oxford History of English Literature was adopted in 1935 the Delegates of the Press invited F. P. Wilson to share the general editorship with Bonamy Dobrée. Their partnership continued until his death. Wilson was concerned especially with the earlier periods, and the prefaces of many of the published volumes speak warmly of the close and friendly attention that he gave to each book as it was written—'not only a patient and helpful adviser but an active co-worker to a degree quite beyond what might be expected of a general editor'; 'such painstaking and skilled help as few authors have ever had from their friends'. Such generosity with his learning and his time was characteristic of the man in all he did. His frank geniality and depth of understanding gave encouragement and inspiration to every younger scholar who had the good fortune to know him.

This lavish giving of time and energy stood in the way of his own writing, but it meant much to the Oxford History. He was to have written the history of the English secular drama to 1642. When he died he had effectively completed the text here presented, and two later chapters which stand somewhat apart and are to be published separately. Though he wrote much else of distinction, this would have been his great work. Now his memorial is to be found in his masterly shorter studies and in the books of other men.

# EDITOR'S NOTE

WHAT is presented here is substantially the work that F. P. Wilson wrote (with the exceptions noted above). My business has been to decide between variant versions, and occasionally to revise what was marked for revision; but I have neither augmented nor curtailed. The Chronological Table, the Bibliography, and the Index are, however, my own work. I must thank the General Editors and Professor Dame Helen Gardner for counsel and help at every stage. I must also thank Mr.

J. C. Maxwell, Professors Werner Habicht and W. A. Armstrong, and Dr. T. W. Craik for advice on the Bibliography. Two secretaries, Mrs. Amphlett and Miss Griffin, have endured much in the cause, and I am grateful for their endurance.

G. K. H.

# CONTENTS

# I

# THE EARLIER TUDOR MORALITY AND INTERLUDE

## 1. *Medwall and before Medwall*

IN the last decade of the fifteenth century John Morton, Archbishop of Canterbury and Chancellor of England, was nearing the end of a long, troubled, and distinguished career. We remember him from Shakespeare's *Richard III* as the Bishop of Ely who was no party to Richard's schemes and in the gardens of whose house in Holborn grew 'good strawberries'; but since the accession of Henry VII, whose cause he had served so well, many honours had been showered upon him. In the household of this prince of church and state, so Roper tells us, the young Thomas More had served as a page, and at Christmas revels would 'sodenly sometymes stepp in among the players, and never studyinge for the matter, make a parte of his owne there presently amonge them' to the delight of his old master. More was later to pay tribute to Morton's memory in a famous passage in the *Utopia* and to record the pleasure this old man took to test, many times with 'roughe speache', the promptness of wit and boldness of spirit of those who came before him. While More was still a page, Morton had shown his judgement by recommending to Henry that early humanist, Adrian of Castello, 'which firste of our tyme' wrote Hall, 'after that golden worlde of Tully, . . . taught the trade and phrase to speake fyne, pure, freshe and cleane latyn'. Morton's household at Lambeth would not be less glorious after he had been created Cardinal in 1493, but we lack particulars. Yet we may adduce the evidence from the Household Book of his young contemporary the fifth Earl of Northumberland to illustrate how richly enterprising in matters of entertainment the great households of those days could be. Rates of payment were fixed in the Earl's household for players of other earls that paid visits between Christmas and Candlemas and also for the domestic players: the children of his chapel playing the Nativity play in the Earl's chapel, or the Shrove Tuesday play, or the play of the Resurrection. A special servant was ordained master of the revels with the task of overseeing plays and interludes during the twelve days of Christmas, while of the six chaplains in household one was the 'Aumer,[1] Ande if he be a maker of Interludes than he to have a Servaunte to th' entente for writynge of the parties And ellis to have noon'.

One of the chaplains of Cardinal Morton, whose business it was to write interludes for the instruction and entertainment of the household and its many visitors, was Henry Medwall.[2] No doubt Medwall's dramatic entertainments included miracle plays as well as prologues for pageants, but the only two plays which have survived are *Nature* and *Fulgens and Lucrece*. A third play, *The Finding of Troth*, which is said to have wearied Henry VII, depends solely upon the word of J. P. Collier.

*Nature*, upon which alone older histories of the drama were able to base their estimate of Medwall, is a characteristic but by no means outstanding example of a fifteenth-century morality play. These plays have been considered, together with miracle plays, in another volume in this series,[3] but a brief retrospect is necessary if *Nature* is to be seen in its historical setting.

The great cycles of miracle plays, associated with such towns as York, Chester, Coventry, Wakefield, and Lincoln, had taken shape long before the year 1485. Recent research is inclined to assign to the last quarter rather than the early years of the fourteenth century, the shift from the liturgical plays, acted by the clergy in churches, to the miracle plays, acted by craft guilds in the streets, with the consequential shift from Latin to English. Even so, by 1485 these plays had behind them a century of tradition, and while a form of drama that depended so much on changing local conditions could never take final shape and often required adaptation and enlargement, yet the manner of it was set by the early fifteenth century when

[1] *Aumer*, almoner.

[2] Little is known of him. According to John Pits he was born in England of a noble family and wrote many works most of which have perished. He was ordained acolyte in April 1490, but does not seem to have proceeded further in holy orders. Various livings were granted to him by the Crown, and he is described in Morton's Register as 'Capellanus'. He resigned from his Calais living in 1501, and perhaps did not long survive Morton's death in October 1500. [See further in Bibliography.]

[3] E. K. Chambers, *English Literature at the Close of the Middle Ages*, OHEL.

the anonymous Wakefield Master is believed to have made his remarkable additions. Much in these plays, then, dates back to the hey-day of English medieval poetry. The craft guilds were responsible for the acting of them and for the elaborate organization necessary to produce them, together with the not inconsiderable expenses; and they did all this with the blessing of the church, for in the streets and in English the plays presented the doctrine of the Church as faithfully as those acted in churches and in Latin. What survives has come down to us in degraded texts, texts which may be as corrupt as those of many a ballad; yet though sometimes pedantic, sometimes dull and crude, they often interpret the drama of the Christian religion with a moving simplicity which has triumphed over all the mutations of taste and belief. Nor do they find humour incompatible with religion. They enhance tragedy with comedy long before the Elizabethans, and if they lack historical sense they possess every other kind of sense. These plays—and not only the cycles but the plays out of cycle, biblical and hagiological—can only receive incidental mention in this volume, but let it not be forgotten that they continued to be performed with unabated popularity until the Reformation and even beyond. England had popular stages long before the Globe and the Fortune, stages with elaborate scenery and costume. Those stages Shakespeare might have seen if he had visited Coventry before 1580 or thereabouts, but not after. The hostility of Elizabeth's ecclesiastical authorities to plays which they condemned as Papistical, the Protestantism of many in the audience, the rival claims of other kinds of drama taken to the people by increasingly popular and numerous bands of professional actors, brought the craft cycles to an end before the close of the sixteenth century. In Spain miracle plays and moralities developed into the *comedias divinas* and *autos sacramentales* of Lope de Vega and Calderon; and (who knows?) if England had remained Catholic perhaps Shakespeare might have written religious plays for street performance on Corpus Christi day.

The miracle plays were developed from antiphonal elements of the Mass and were one means of teaching sacred history, whether of Christ or His saints, at a popular level; morality plays may have developed from the homily and were one means of teaching Church doctrine, again at a popular level. In early years both kinds may have been written by secular clergy in close touch with the people. Four great cycles of miracle plays have survived, and some plays out of cycle, but the remains of early morality drama are exceedingly scanty. Till recently the Pater Noster play at York, to which Wycliffe referred in 1378, has been taken to be a morality play and the first of which we have mention, but Professor Hardin Craig has shown how doubtful this assumption is. The play was in the charge of the guild of the Lord's Prayer, its purpose was to teach the Prayer, and in setting forth its goodness, says a guild document of 1389, 'all manner of vices and sins were held up to scorn, and the virtues were held up to praise'. Another guild document of 1399 mentions the play of 'Accidie' (Sloth), and this may be linked with the tradition that the seven petitions of the Prayer were each a means of salvation from one of the seven deadly sins. The play, therefore, may have been in seven parts (or eight if there was an Induction), acted processionally; other evidence suggests that each part may have illustrated the victory over a deadly sin not in morality form but from the life of a saint. If so, Tarlton's *Seven Deadly Sins* bears a striking resemblance in form, though his exempla are secular. Part II of his play (*c.* 1585), a skeleton outline of which has survived, illustrates the sins of Envy, Sloth, and Lechery from the stories of Ferrex and Porrex, Sardanapalus, and Tereus. The Pater Noster play survived the Reformation, for there were performances in 1558 and 1572, but in 1575 it did not survive the scrutiny of Archbishop Grindal.

The earliest morality texts that survive are *The Castle of Perseverance*[1] and the fragmentary *The Pride of Life*. They belong to the early years of the fifteenth century, and they are among its most impressive works. *The Castle of Perseverance* (*c.* 1425) comprises in its massive structure all the leading themes which recur singly in other moralities but not elsewhere together. The play traces the course of Mankind's life from youth to age, and represents the conflict for his soul between the virtues and the vices. His Bad Angel brings him the World, the Flesh, and the Devil, and the seven deadly sins; his Good Angel sends him Conscience, Confession, Penance, and the seven cardinal virtues. These call him to a castle stronger than any in France, the Castle of Perseverance. Mortal life is shown as a battle or a siege in which the prize is the soul of man and the combatants

[1] *Perseverance*, steadfastness, constancy. The accent was on the second syllable.

the virtues and vices in his soul. It is for man to choose salvation or damnation with 'fre arbritracion', but because the good and evil impulses of his nature are externalized in allegorical personifications, there is dramatically no civil war within the soul, as there is in some kinds of tragedy. This conflict between virtue and vice is the leading theme in *Wisdom* (*c.* 1460), a play devotional and doctrinal, where the struggle is between Wisdom (Christ) and Lucifer for man's soul (Anima) and its three powers, Mind, Will, and Understanding. It is also the leading theme in *Mankind* (*c.* 1471), a play in which the comic scenes are vulgar but not funny and almost swamp the moral scenes. If this play was written by a priest, perhaps he was a hedge priest.

With this conception of the soul of man garrisoned in a frail carcase, defended by the powers of good and besieged on all sides by the powers of evil, the nobly serious *Castle* combines the solemn theme of the coming of Death. 'Evyr at the begynnynge Thynke on youre last endynge.' What in the *Castle* is one of several themes is the main theme in the fragmentary *The Pride of Life* (of about the same date), and in the justly famous *Everyman* (translated from the Dutch, *c.* 1500): the certainty of death and the reluctance of man to prepare for his end. In the one, 'drery Dethe' overcomes a king who has put his trust in two knights, Might and Health: in the other, he overcomes representative man. What makes *Everyman* the most impressive of the early remains of this kind is the concentration upon this limited theme, together with the dramatic aptness of the allegory, the sombre devotional treatment, and the choice keeping of the diction.

Yet a third theme is made use of in the *Castle*. When Mankind dies in his sins, he puts himself on God's mercy, for only Mercy can save his soul (Anima) from hell. There follows the debate of the Four Daughters of God which the Church had developed from Psalm lxxxv. 10: 'Mercy and truth are met together: righteousness and peace have kissed each other.' These four characters argue before God's throne; by the clemency of God the pleas of Mercy and Peace are triumphant, and man is saved. The theme is found also in the 'Salutation and Conception' of the N-town (formerly called the Coventry and now believed by some to be the Lincoln) miracle plays and in a miracle play (*Processus Satanae*) of which we possess only the part of the actor who played God; it is touched on in *Mankind*, and it is used for purposes political rather than devotional in *Respublica*.[1]

Medwall's *Nature*, like the *Castle*, traces the course of man's life from youth to age and represents mortal life as a battle between the virtues and the vices for the possession of man. Nature, after whom the play is called, presides only over the opening scene. She is the 'Worldly goddess' appointed by God to knit all living things together and maintain them in their degree. It is she

> Who taught the cok hys watche howres to observe
> And syng of corage wyth shyrll throte on hye
> Who taught the pellycan her tender hart to carve
> For she nolde suffer her byrdys to dye
> Who taught the nyghtyngall to recorde besyly
> Her strange entunys in sylence of the nyght.

But unlike the *Castle* the play does not survey the life of man from the womb to the tomb and beyond to the Doom: it stops short of the coming of Death. When Man enters into the World (Mundus), he disobeys Nature's counsel to subordinate Sensuality to Reason and Innocence, becomes subject to Worldly Affection and the seven deadly sins, repents, then relapses, and in old age is reconciled once more to Reason. Medwall must have heard this orthodox doctrine preached in many a sermon before he presented it in dramatic form, and here he strikes a nice balance between *sentence* and *solace*, the *solace* or comic and satirical strokes in the play being assigned as ever to the vices.

The *Castle* is an outdoor play with an elaborate setting: *Nature*, as we should expect, bears every sign of having been acted indoors, most probably in a hall in Morton's palace by the servants of the household under the direction of the author himself. This and Medwall's *Fulgens and Lucrece* are among the very few interludes divided into two parts, each of about 1400 lines, the division in *Nature* coming after Man's first repentance. The audience is warned that much more of the 'processe' will be shown to them 'when my lord shall so devyse'. So each half of the play approximates to the length of the short interlude of the earlier sixteenth century.

[1] Below, pp. 40 ff.

As the author of *Nature* Medwall would barely merit a mention in the history of our drama, but as the author of *Fulgens and Lucrece*, the first purely secular English play that has survived, he is a significant figure. We have to wait for many years before meeting with another play so free from religious and allegorical treatment and so original in its method of adding pleasure to instruction. The discovery of this play in 1919 caused almost as much surprise as the recent discovery of fragments of an historical play dating from the great period of Greek drama.

No doubt Medwall's play seems to us more original than in fact it was. The remains of secular drama before his day are exceedingly scanty, yet they are sufficient to show that miracle and morality plays do not cover the whole range of dramatic entertainment in the later Middle Ages. In addition to what we know or may surmise of the activities of minstrels and travelling *joculatores*, of May games, Whitsun pastorals, and other folk-festivals, of mummings, of Christmas Lords, of elaborate and costly pageantry in which mythological, historical, and even secular scenes were interpreted by prologues from the pens of Lydgate, Medwall, and others,[1] there is the remarkable fragment *Interludium de Clerico et Puella* of the late thirteenth century, a dramatization of a *fabliau* in which a priest, rejected by the maiden he has wooed, seeks the aid of a bawd. There is the actor's part for the fourteenth-century *Dux Moraud*, based on a very different sort of *fabliau*, with its themes of incest and repentance. There are fifteenth-century references to plays based on popular romance themes—Robert of Sicily, Sir Eglamour of Artois, and Le Bone Florence of Rome. There are traces of a St. George play from 1456; and there is a chance reference by Sir John Paston in April 1473 to an unsatisfactory servant, as one whom he had kept 'thys iij. yer to pleye Seynt Jorge and Robyn Hod and the Sheryff off Notyngham'; what has been taken to be a fragment of such a play—the plot that of the ballad 'Sir Guy of Gisborne'—has survived in a manuscript of about the same date. Add two popular plays, more or less imperfect, which got into print *c.* 1560, the one on 'Robin Hood and Friar Tuck' and the other on 'Robin Hood and the Potter', 'verye proper to be played in Maye games', and then add Sir Edmund Chambers's comment on these fragmentary remains: 'So we build up the past.'

[1] Below, pp. 80 f.

Medwall's *Fulgens and Lucrece* has no relation to *fabliau* or to medieval romance or to popular hero-worship. He takes a favourite medieval form, the *débat*, and gives it dramatic shape; he takes also a favourite topic, as popular in the fifteenth century as in the sixteenth, whether gentility resides in birth or in worth. Some twenty years later it was to furnish either John Rastell or John Heywood with the theme of his interlude *Gentleness and Nobility*.[1] In Medwall's play Fulgens, a senator of Rome, has a daughter Lucrece who is wooed by a wealthy, well-born, but profligate suitor Cornelius and by Gaius Flaminius who by his services to the state has risen from low degree to great honour. Given the liberty of choice—and here Medwall departs from his source—Lucrece after hearing speeches from both suitors chooses Flaminius, while urging the audience not to take her words 'by a sinistre way' or to imagine that she despises the 'blode' of Cornelius.

Medwall's source has been identified as a prose *Controversia de nobilitate* written in 1428 in the form of a *novella* and in humanistic Latin by Buonaccorso da Montemagno. But although Medwall was chaplain to a churchman who knew the value of humanistic Latin in the modern world, we must resist the temptation to call him a humanist. He depended not on Buonaccorso's Latin but on the translation made by the great John Tiptoft, Earl of Worcester and printed by Caxton in 1481; and Tiptoft himself, though a patron with humanistic leanings, may have translated from a French translation. Humanism is a word to avoid in any consideration of early English drama, for if it means the ability to write humanistic Latin, our dramatists were much better occupied in making their own language more expressive; and if it means the recovery of Greek, they never recovered Greek, and even after readers of English came to know Sidney's *Apology* with its Italian interpretation of Aristotle's *Poetics*, all but a few dramatists ignored every precept they were advised to follow.

*Fulgens and Lucrece*, then, derives from an English translation of a work treating a favourite medieval theme in the favourite medieval manner of a *débat*. It presents a secular theme, and for the first time a woman character in an important role. But more striking than the secular character of the play and the absence of allegory is the introduction of a sub-plot

[1] Below, pp. 25–26.

attached to the main plot and illuminating it at a comic level. Two characters A and B push their way through the audience and discuss the play about to be presented. They also push their way into the play—as Thomas More is reported to have done in Morton's household[1]—A taking service with Flaminius and B with Cornelius. To them, to their cross-wooing of Lucrece's clever maid, and to her discomfiture of them both are given nearly two-thirds of the first part, and their share in the second part is also substantial. They keep the audience merry by mis-carrying messages, misprising the meanings of words, making satirical comment on women and on extravagance in costume, boasting of a courage which they do not possess and of feats which they do not perform, and by many a grotesque expression and obscene innuendo. If much of the comedy is coarse—and it was long before the physical results of fear ceased to raise a laugh—it passed for 'right honest solace' in a cardinal's household, especially in time of festival. There is hardly a jest which cannot be found in later plays, and no doubt many had already met with acceptance at earlier feasts and at folk festivals. What seems to us so original is not merely the addition of a sub-plot but of a sub-plot that parodies the serious plot. The wooing of the maid gives the obverse of the wooing of Lucrece, and sentiments like 'He is well at ease that hath a wyf, Yet he is better that hath none be my lyf' and 'He that hathe moste nobles in store Hym call I the most noble evermore' are the earthy accompaniments to the sage and serious comments of Lucrece. In short, the rudimentary A and B may dimly recall the effects of Falstaff's presence in *Henry IV* and of Touchstone's in *As You Like It*.

What induced Medwall to elaborate his sub-plot was first the necessity of entertaining his audience, a seated audience which had already eaten and drunk its fill. A observes that while some members of the audience delighted in serious matter 'be it never so longe', others cared for nothing but trifles and jests; and as it was the author's intention to please 'the leste that stondyth here', 'dyvers toyes' had to be introduced, even if impertinent to the principal matter. On the same principle Sir Thomas More allowed the preacher to arrest the attention of his congregation with jests, for if a man will not listen to talk about heaven unless he is refreshed from time to

[1] Above, p. 1.

time with a merry foolish tale, 'there is none other remedy but you must let him have it'. Another consideration was the necessity of writing a play in two parts and the knowledge that the serious plot alone could not sustain more elaboration than it had already received. When the play opens the spectators are seated at a feast, a feast which is renewed at the end of Part I at the command of 'the master of the fest'. Part II, the 'remnant' of the play, was acted after dinner on the evening of the same day 'aboute suppere'. The hall in which the actors performed was doubtless Morton's banqueting hall, and the mention of a fire in the hall suggests a winter season.

Both *Nature* and *Fulgens and Lucrece* are entitled 'goodly (godely) interludes', and there has been much speculation whether the word 'interlude' is by derivation a play (*ludus*) conducted between (*inter*) two or more actors or a play performed between the courses of a banquet. The latter explanation applies more aptly to the early Tudor interlude which is distinguished from earlier morality plays by its brevity, a smaller cast of actors (often four men and a boy), and an absence of scenery. That there was interval enough for plays and other entertainment during the gargantuan feasts of the fifteenth and sixteenth centuries is suggested by Robert Lindesay of Pitscottie's description of the triumph and banquet given by James IV at Holyrood in or about the year 1505. It lasted three days from 9 a.m. till 9 p.m., and 'betuix everie service thair was ane phairs[1] or ane play sum be speikin sum be craft of Igramancie'. It is impossible, however, to tie the word 'interlude' to any one meaning, for in one place or another it is found attached to every kind of drama known to the Middle Ages. Even in Tudor times Skelton's *Magnificence* was considered as much a 'goodly interlude' as either of Medwall's plays, so that brevity is no distinguishing mark. In a lawsuit in which John Rastell was involved, a clear distinction was made between open-air 'stage plays in the sommer and interludes in the winter', the cost of hiring costumes for the outdoor stage-plays being several times greater than that for the indoor interludes; but the distinction is not found elsewhere unless it be in a royal proclamation of 16 May, 1559. There it is stated that the time of interludes in the English tongue was past till

[1] *phairs*, farce; but farce apparently included conjuring and necromancy ('Igramancie').

the next Allhallowtide (1 November). The word lasted on into the seventeenth century, especially in official documents, long after changes in dramatic taste and the use of preciser (but still imprecise) labels—comedy, tragedy, chronicle history, pastoral—had rendered it superfluous. To the historian it still has its uses, and when it is used in this book it refers to the short indoor Tudor play of about a thousand lines, whether allegorical or not allegorical.

## 2. *The Two 'Makers': Skelton and Lindsay*

Although brevity is one of the distinguishing marks of the early sixteenth-century interlude, the honour due to poetical merit demands that full consideration be given to two important exceptions, one Tudor and one Stuart: John Skelton's *Magnificence* and Sir David Lindsay's *Satire of the Three Estates.* The historian of early sixteenth-century drama may well pause at these names, for here are two known poets, and good poets, who have turned to drama. We cannot say that of any men before them, and we have to wait many a year, perhaps till the time of George Peele, before we can say the like again. Skelton[1] is no longer the earliest English dramatist to whom we can assign a name and an identity—that distinction must go at present to Henry Medwall—but he is the first poet to use drama to express a view of life which may be amply illustrated from other works of his that have survived. As a dramatist he can be judged solely by *Magnificence.* His 'soverayne enterlude of Vertu' and his 'comedy called Achedemios', probably a school drama, have perished. The recently discovered fragment of a morality (printed by William Rastell in 1533) containing a character called Good Order has been identified by some with the *De bono Ordine* ascribed to him by Bale. The doctrine is his, for this disorderly man believed above all things in Good Order;

[1] Skelton (1460?–1529) was probably at Cambridge before receiving the degree of poet laureate at Oxford to which Caxton referred in 1489. Later, he was given similar degrees (for proficiency in rhetoric?) by Louvain and Cambridge. His earliest extant poem relates to the murder of Northumberland in 1489. Tutor to Prince Henry until Henry became heir to the throne (1502), he was ordained priest in 1498 and by 1504 was Rector of Diss, Norfolk. He appears not to have resided there after 1511, but to have lived in Westminster, from 1512 using the title *orator regius.* His satires on Wolsey belong to the years 1521 and 1522, and at this time he lived in sanctuary at Westminster. In 1523, with *The Garland of Laurel,* he made his peace. He was buried in the chancel of St. Margaret's, *vates Pierius.*

but, then, any dramatist of his age might have shown Old Christmas beguiled by Riot and Gluttony, rescued from their clutches by Good Order and the action moralized by Prayer and (possibly) Abstinence. There is nothing distinctively Skeltonic in the writing, and the resolve of the banished Gluttony to depart for 'the new founde land. . . . For in England there is no remedy' would provide, if it were Skelton's, the only indication in his works that he knew himself to be living in an Age of Discovery. His *Necromancer*, alleged by Thomas Warton to have been printed by Wynkyn de Worde in 1504 as played before Henry VII at Woodstock, and therefore a few years earlier than Ariosto's *Negromante*, represented the trial of Simony and Avarice with the devil as judge, but before we lament the loss of this interlude we have to decide whether it is utterly incredible (as Joseph Ritson maintained) that it ever existed and whether, like passages as circumstantial, it was not fabricated by the first historian of our poetry.

So Skelton stands or falls as a dramatist by *Magnificence*. The play is of some length (2,567 lines), and, while undivided into acts, the action falls naturally into five parts named by the learned editor, R. L. Ramsay, Prosperity, Conspiracy, Delusion, Overthrow, Restoration. Magnificence represents the power of a man of high estate to use position and wealth (Felicity) without losing either and without losing his own soul. Unless controlled by Measure and Circumspection man's reason will be subject to his will (or desires), Liberty will become licence, and Felicity will be lost to him. 'Measure is treasure', and all is well while Magnificence is content to subordinate Felicity and Liberty to Measure's control. But suborned by Fancy (fantastic wilfulness), Magnificence banishes Measure from his counsels and becomes a prey to Fancy's crew of courtly companions, Counterfeit Countenance, Crafty Conveyance, Cloaked Collusion, and Courtly Abusion. Last of all comes Fancy's brother, Folly, to complete the ruin of Magnificence. Felicity lost, Adversity and Poverty take charge, and he is saved from Despair and Mischief only by Good Hope. His restoration is completed by Redress, Circumspection, and Perseverance, and he returns to his palace with joy and royalty.

The moral is clear and insistent, and the high seriousness of the theme is conveyed towards the end in the speeches of Adversity and of the three Virtues:

This treatyse, devysyd to make you dysporte,
Shewyth nowe adayes howe the worlde comberyd is,
To the pythe of the mater who lyst to resorte:
To day it is well, to morowe it is all amysse;
To day in delyte, to morowe bare of blysse;
To day a lorde, to morowe ly in the duste:
Thus in this worlde there is no erthly truste.

It is a solemn music[1] and for the moment we are taken back in spirit to earlier morality plays, the sole concern of which is with the salvation of mankind. Like the *Castle of Perseverance*, *Magnificence* represents a Conflict between Virtues and Vices with an admixture of neutral characters like Liberty—

For I am a vertue yf I be well used,
And I am a vyce where I am abused.

But it is only at the end—during Magnificence's overthrow and restoration—that the play takes this wide sweep. Magnificence is not Mankind but 'a prynce of great myght', and the play is not so much a mirror for man as a mirror for princes, a lesson in the art of good government, but also a lesson in the art of preserving worldly prosperity. Was it addressed like Skelton's earlier moral treatise *Speculum principis* to his old pupil Henry? The application could not fail to have been made. From internal evidence the play has been assigned to the year 1516, by which year the rich treasury inherited by Henry from his sober father had been dissipated on his own accomplishments and pleasures and by Wolsey on a disastrous foreign policy. Was Skelton's play an attempt to call a halt, and was Wolsey one of the evil counsellors at whom the dramatist points? In a play written for performance (apparently at night: cf. l. 365), particular reference would be well disguised and the poet is much more successful than he was in *Speak Parrot* in making allegory 'his protectyon, his pavys,[2] and his wall'. Yet like the Wakefield Master a century before, Skelton was a man who found it difficult not to write satire, and if his play was ever presented at Court, Magnificence himself might have winced at some of Skelton's gibes, and many a courtier might have identified Courtly Abusion and the other crew of vices not with himself perhaps, yet with many another courtier.

[1] Rhyme royal: a b a b b c c.

[2] *pavys*, shield.

Some of Skelton's greatest admirers have found *Magnificence* disappointing and have preferred to write about its historical interest. Every work that Skelton wrote has some pith in it, but pith is not easily sustained through so long a work. It has been well said that Skelton's poems grew by accretion, and his best work is done in kinds of poetry that need no predestined shape. The length of the speeches in which both virtues and vices introduce themselves, good as some of these are, the lack of any nice discrimination between the vices, the elaboration of the 'comic' scenes, often in rapid stichomythic dialogue which in time becomes tedious to the ear, the failure (inevitable to the allegorical drama) to present on the stage the conflict between Virtue and Vice in the mind of Magnificence—these detract from the dramatic interest. But the historical importance of the play is considerable. We have here, perhaps for the first time in English drama, a work of conscious art, a laureate devising a long work in conscious observance of rhetorical principle. This appears not only in the crafty use of the 'polysshed and ornate termes' for which Caxton praised him but in the versification, also a part of rhetoric. The dignified rhyme royal is reserved mainly for the Virtues and for Magnificence. The long-lined couplet is common ground: the sentiment and the diction mark the distinction. Lighter forms of verse are reserved for the vices—whether the light four-stressed couplet or that dancing two-stressed line prized by lovers of Skelton's lyrics. And in the lighter forms of verse the poet may write in couplets or he may let loose a 'leash' of rhymes, the same rhymes continuing until they appear to be exhausted, 'bastarde ryme, after the dogrell gyse' as Fancy calls it, a verse which Skelton has so made his own that no later poet has been able to use it without reminding us of its begetter.

Here then is a conscious attempt to introduce variety of verse and to fit the verse to the character. It has been said that Chaucer proceeds from complex metre to simple, and from simple matter to complex. The drama, too, as it moves from the simplicities of *Magnificence* to the complexities of *Hamlet*, proceeds from complex metre to simple. Skelton has abandoned the intricate stanza forms of the miracle play, but he is still experimenting with a great variety of verse. Later in the century, as we enter the Great Age, even rhyme tends to disappear and we are left with blank verse and prose.

But the difficulty in reading Skelton lies not in the variety of the verse but in the prosody of the long line, a difficulty which assails any modern ear in reading almost any formal verse written between Chaucer and Surrey, and almost any dramatic verse before *Gorboduc*. One is tempted to apply to much of this verse the line in which Ariosto (or rather Harington) describes the horsemanship of the hermit in *Orlando*, viii. 26: 'His trot was very bad, his gallop worse.' In the lyric and in the lighter forms of verse we are not troubled, but how are we to speak the full heavy verse of which a stanza has been quoted on p. 13? The 'tumbling' verse of the fifteenth and sixteenth centuries was marked by the presence of rhyme and four stresses with a caesura normally after the second stress, and we deprive these lines of their due if we attempt to scan them as 'lame pentameters'.[1] Yet an ear tuned to the niceties of Chaucerian verse or the regularities of later dramatic verse fumbles with the stresses, and dithers between rival systems of versification. We may prefer Skelton's uncertain stresses to the monotonous certainties which blighted dramatic verse in some early Elizabethan plays, but we may be thankful that before the century was out the poets found a law of metre within which they could permit themselves an infinite liberty of rhythm.

If *Magnificence* differs from earlier moralities by its more restricted theme and more secular tone, it differs from the interludes then becoming the vogue by its length and elaboration. The characteristic interlude of early Tudor times shrinks to the dimension of a thousand lines. Yet *Magnificence* is continence itself compared with Sir David Lindsay's *Satire of the Three Estates*.[2] The play was first produced at Linlithgow on 6 January (Twelfth Night) 1540 before James V and his Queen, Mary of Lorraine, and the whole Council, spiritual and temporal. This text has not survived, but an eyewitness's summary of the action suggests that it was an indoor interlude (the date ensures that), with fewer characters than the later

[1] Below, p. 68.

[2] Lindsay (*c.* 1486–1555), eldest son of David Lindsay of the Mount, near Cupar, Fifeshire, received a pension from James IV and a playcoat of blue and yellow taffety in 1511, after Flodden became usher (but not tutor) to the infant James V, was a herald by 1530, and by 1542 had become Lyon King of Arms and a knight. He travelled extensively in England, France, Denmark, and the Low Countries on various missions. None of his extant poems may be dated earlier than 1528, and his earliest appearance as a reformer is in his *Complaint* of 1529. In spite of his outspoken satire, he retained his office as herald until his death.

texts and less comedy; it was, however, as outspoken in its exposure of social evils and especially of the corruption of the spiritualty. The suggestion has been made that it was propaganda for the reform party at the Scottish court and an attempt to persuade the king to break the alliance with France and to seize Church property after the example of his uncle in England. The play, so our informant tells us, incited an angry king to threaten his bishops that if they did not reform he would send some six of the proudest of them to England; but it did not induce James to throw in his lot with a Protestant England, and his death in 1542 was hastened by the shame of Solway Moss. This first draft was drastically revised and expanded for an open-air performance on 7 June, 1552 at Cupar, Lindsay's home ground, on the hill where once had stood the castle of Macduff. This second version has been preserved in part by that great benefactor of Scotch poetry, George Bannatyne, in the manuscript of 1568 which bears his name. The quarto of 1602, published in Edinburgh by Robert Charteris, gives the text performed in that city at the Playfield on the Greenside, Calton Hill, on 12 August 1554 before the Queen Regent and many of the nobility. Bannatyne, being more interested in the 'merry Interludes', omitted much grave matter on the ground that the abuses attacked were by 1568 well reformed, so that he preserved only 3038 lines of the 4630 in the quarto. The only important lines preserved by Bannatyne and not in the printed text are the 'banns' or preliminary announcement of the time and hour of the Cupar performance. These give no indication of the sharp and serious satire to come, but whet the appetite of the audience by showing the humours of a cottar and a shrewish wife, an old man with a young wife, and a *miles gloriosus* who is discomfited by a fool, three time-honoured themes. One does not need to be a Zeal-of-the-land Busy to find the episode of the old man with a young wife very coarse, exceeding coarse, very exceeding coarse. One does not need to be a Lucio to decide that, if manners and morals permit this sort of low comedy, Lindsay does it with a race and wit beyond the reach of the Englishman who wrote *Mankind.*

The length of the play puts it in a class by itself. It dwarfs *Magnificence* and still more so the short-breathed interludes of early Tudor England. The Cupar banns warn the audience to attend 'richt airly' at 7 a.m.; at the end of Part 1, timed by the banns for 11 a.m., the audience is told to 'tak ane drink and mak Collatioun' but not to tarry long, it being 'lait in the day', and the best part of the play yet to come. And the second part is longer than the first. At Edinburgh, according to Henry Charteris's preface to the *Works* (1568), the performance lasted from 9 a.m. till 6 p.m. Why it should have taken so long is mysterious, for Lindsay's verse moves swiftly, and there are more lines with four or fewer stresses than with five. But time had to be allowed for a 'Collatioun', for processions, for stage business, and for laughter. The stage, too, was large and, like the play, panoramic. At Edinburgh a scaffold high enough to demand a ladder was built for Rex Humanitas and his throne, and no doubt the action of the play was made clearer by the use of multiple settings, each character or set of characters being provided with its own 'mansion'. A chamber to which the king retired and a pavilion for other characters were required, and a 'house' or pavilion from which the Queen Regent watched the play. Properties were elaborate: among them a pulpit, a bar to plead from, and stocks and a gallows each roomy enough to accommodate three characters at a time. Costumes, perhaps, were not gorgeous, but the town-painter received £5 for painting the scenery and the players' faces. Nature provided the hill at the back from which the pardoner's boy shouts his discovery of a great horse-bone, the bone of St. Bride's cow, and a practicable stream or ditch near the front or side of the stage over or through which characters leap or wade and into which Pauper pitches the Pardoner's relics. For any English parallel to this elaborate setting we should have to go back a century and a half to *The Castle of Perseverance*. The ground-plan of the stage for that play so providentially supplied in the Macro manuscript shows an arena surrounded by a ditch or fence with the castle in the centre and five scaffolds at the circumference. It has been argued that the stage for *The Three Estates* was circular with the audience disposed in just over a semi-circle; but there is no evidence. (In the brilliantly successful revival at Edinburgh in 1948 the play was produced in a square galleried chamber on a platform stage open on three sides.) Of one thing we may be sure: there was great intimacy between the actors and the audience. Some of the actors appear to pass through the audience on their way to and from the stage, and there is much direct address. Historians of the drama have often taken direct address as evidence of crudity, but Lindsay, like Medwall and others before him, was wise in his generation, and one reason why he sustained the interest of 'ane exceeding greit nowmer of pepill' for so many hours may be that he set up no barriers but brought his audience into his play. Properly handled, no audience objects to participation.

We may marvel, with Henry Charteris the publisher, how Lindsay dared so plainly to inveigh against the vices of all men, but chiefly of the spiritual estate, and that at a time when 'they had the ball at thair fute'. A king, Rex Humanitas, his mind at first *tanquam tabula rasa* and as ready for good as for ill, is tempted by three evil counsellors, Wantonness, Placebo, and Solace, and falls a prey to Sensuality. To their number are added three vices, Flattery, Falsehood, and Deceit, who give the dramatist the chance of satirizing the three estates (Lords Spiritual, Lords Temporal, and Burgesses). They refuse Good Counsel (absent from Scotland this many a year) access to the king, and encouraged by the spiritualty place in the stocks Verity, who arrives with the New Testament in English, and Chastity, who gets no help from any class of society. At the end of Part I Divine Correction arrives in Scotland with the intention of reforming the three estates. In terror the three vices take shelter with the Church, the merchants, and the craftsmen; Wantonness, Placebo, and Solace are pardoned on promise of reform, because a king needs innocent amusements; and Good Counsel, Verity, and Chastity force Sensuality to take shelter in the Church, and are admitted into the king's service. A summary does no justice to the sharp satire in the speeches of the vices or to such low comedy as the rejection of Chastity by the wives of a souter[1] and a tailor. In Part II the satire is directed more particularly at the spiritualty. This part begins with the only passage in the play which may be truly called an interlude. Before the king and the principal players have taken their seats a poor man (Pauper) inveighs against the rapacity and immorality of the clergy, and a pardoner exposes his deceits and his relics. At a comic level the interlude prepares the audience for what is to come. The parliament of the three estates, summoned by Diligence at the end of Part I, are led to their seats by their vices. If they walk backwards, it is because they have done so this many a year

[1] *souter*, cobbler.

and like it. Two new vices make their appearance: Covetise (the spiritualty) and Public Oppression (the temporalty), while Falsehood and Deceit still wait upon merchants and craftsmen, and Flattery upon all. John the Commonwealth and Pauper present the grievances of the common people and spare no class of society. The Lords Temporal and the Burgesses agree to reform, but the Church, represented by a bishop, an abbot, a prioress, and a parson, plead that they live as they do after old use and wont. Three prudent and devout clerks dispossess the spiritualty, and a model sermon points the moral that the clergy should not delegate this duty to friars. Fifteen acts of the Parliament are proclaimed, the last ten of which are strictures upon the spiritualty. The play is not yet ended, for Common Theft, Deceit, and Falsehood have yet to be hanged, Flattery assisting. And Folly preaches a sermon, in which neither king nor peasant is spared, on the text *Stultorum infinitus est numerus*.

The Scotch lords complained that James V was 'ane better priestes king nor he was thairis'. Lindsay attacks abuses in all three estates, and some of his most searching satire is put into the mouths of John the Commonwealth and Pauper; but his main target is the Church. If until the eighteenth century Lindsay's poems were thought little less necessary in every family than the Bible, and his name and fame were remembered when Dunbar and Douglas were forgotten, the reason is that the reformers took him to themselves. It was held that he prepared the ground for John Knox's seed. 'He gave the scarlet dame a box Mair snell than all the pelts of Knox', says Allan Ramsay. Yet we cannot imagine him at home in Knox's world, and there is no evidence that he seceded from the Catholic Church. In the points of doctrine which divided Calvinism from Catholicism he shows no interest: he is for reformation rather than the Reformation, and the reforms for which he presses are not doctrinal, but political and social: the drain of good Scots money to Rome, the excessive mortuary taxes, the oppression of the poor, the folly of war. If he is not of the rank of Dunbar and Douglas, the reason, then, is not David Irvine's (1810), 'the unostentatious genius of the Presbyterian discipline is less congenial to a poetical imagination than the pomp and parade of the Romish superstition'. The absence of formal rhetoric is a negative merit, but a positive demerit in so long a work is his incapacity for song or for lyrical verse of any kind or for solemn music. Yet he is a master of many metres and uses them cunningly for variety. Scotch poetry had not suffered from the prosodic anarchy which beset English poetry; we always know where we are with Lindsay's rhythms, and they seldom fail to be lively and speakable. He shows a nice dramatic instinct in spacing his more serious passages with comic, and he has at command the rich resources of what used to be called Broad Scotch, a language which seems to be made for pungent satire and broad humorous effects. Neither his language nor his allegory suffers from the abstractions which make so much contemporary English work dull and lifeless. Even without the actors to give these allegorical types a human presence, it is easy to see how characters like Sensuality, Flattery, Folly came to life upon the stage, not to speak of the Pardoner, Pauper, the souters and tailors and their wives. What also helps to make his play vivid is the zest for handling a multitude of details, a zest which an acute critic, Gregory Smith, pointed to as one of the distinguishing marks of Scotch poetry. The particular is not necessarily an enemy of the general—and the main lines of the allegory are never confused—but it *is* an enemy of the abstract. If *The Three Estates* held and holds the stage, it is because on one vast canvas Lindsay has given us a Scotch *comédie humaine*. We have nothing like it in England.

Lindsay's play is a lonely document in the history of early Scotch drama, and this English writer will not again find it necessary to cross the border. The dearth of drama in Scotland has been attributed to the same reason that there are no snakes in Iceland, the excessive cold. By the same token an Italian might have denied the possibility of drama in England. Scotland was thinly populated, poor, and remote—'the erse of the warld' in a phrase reported by Pitscottie—yet there is plenty of evidence that it had a taste and a turn for the drama. If there had been no evidence we should have had to invent it, for attempts to show that Lindsay was indebted to French *sotties* are unconvincing, and it is as likely that he was drawing on native traditions. An act of parliament of 1555 put down Robin Hood, Little John, the Abbot of Unreason, Queens of the May, yet these folk-amusements survived. We hear of disguisings and pageants at court revels and occasional payments *pro certis ludis et interludiis*, miracle plays were performed by the craft guilds of various towns, and the *spectacula* of minstrels had their subterranean existence as in England. But where are the plays of yester-year? The only survivals of note are Lindsay's play and the banns or 'cry' for a play or May game (*c.* 1500) entitled 'Ane Little Interlud of the Droichis[1] Part of the Play' lively enough to be attributed to Dunbar. After the Reformation the restraints of the presbyteries and especially of the local kirk sessions became more and more irksome, and the Court, which had more reason than most to feel the truth of the Preacher's 'Woe to thee, O land, when thy king is a child', was either too troubled or too much at odds with the people to be an effective patron of the drama. When some English comedians visited Edinburgh in 1599 James VI forced the ministers to rescind their act forbidding their congregations to attend. If the actors took this as a good omen of favours to come, they were not to be disappointed. But it was in England that James nourished the living; in Scotland he could not resuscitate the dead.

### 3. *Rastell and Heywood*

To consider Lindsay's play immediately after Skelton's is to flout chronology for the purpose of bringing together two famous poets. We must return to England and to the England of *Magnificence*. A glance at three early interludes, *The World and the Child* (*Mundus et Infans*), *Hickscorner*, and *Youth*, all of which belong to the earliest years of the century, will serve to indicate the form from which the normal sixteenth-century morality play was not seriously to depart. None of them exceeds a thousand lines in length or requires more than five or six actors. *The World and the Child* attempts to crowd into its narrow compass the temptation of Manhood in the World from the cradle almost to the grave, until old and broken he is rescued from Folly by Conscience and Perseverance. There is much doctrine set forth systematically—the deadly sins, the five bodily and the five spiritual wits, the twelve Articles of Faith. The author of *Youth* confines himself to Youth, Charity, and Humility rescuing him from Riot, Pride, and Lechery, the interlude rushing to a conclusion with Youth's sudden repentance immediately after he has decided to 'make mery whiles I may'.

[1] *Droichis*, dwarf's.

It has been acted with acceptance in modern times. In *Hickscorner*, to which *Youth* is indebted for some of its comic dialogue, repentance comes as suddenly, but otherwise it is of a different pattern. As it is the first play to derive its title from a vice, so it is the first in which the principal character (Freewill) is a comic character. His boon companions are Imagination and the scoffing Hickscorner, and he and Imagination win salvation with the assistance of Pity, Perseverance, and Contemplation. Hickscorner appears to be beyond redemption. Part of the convention in morality plays, as indeed in all moral allegory, is that (as in life) vices disguise themselves under the names of virtues. So in *The Holy War* the Lord Lasciviousness does much damage under the name of Harmless-mirth, until he and his sons Jolly and Griggish are shown up. In this play the regenerated Imagination is renamed Good Remembrance, but Freewill needs no new name, 'For all that wyll to heven hye' must forsake folly by their own free will. There is some edge to the satire of *Hickscorner*, and Pity's indictment of a corrupt society (in stanzas with the refrain 'Worse was hyt never') is unusually effective. In most of these interludes, if the reader tires of the moral abstractions, he may be refreshed from time to time with some bit of social history—a list of the wines on sale in London, the names of ships sailing to and from Ireland, or of games at dice and cards, the manner of life in Newgate, the tricks of conycatchers. Yet it is a gloomy world they portray in spite of the merriment. 'The weeds overgrow the corn' and 'Worse was hyt never'.

All the English plays mentioned so far to which an author's name can be attached were written by men of the church, and for all we know so were the anonymous plays whose titles have been cited. Yet the fifteenth century had seen a marked increase in the literacy of the laity, and for many years learning had ceased to be peculiar to the clergy. William Cornish or Cornyshe (d. 1523),[1] if fate had been kinder, might have appeared in this history as one of the earliest laymen to make important contributions to the drama. As a musician and composer he was a favourite of Henry VIII, and for years he prepared pageants, interludes, devices, and masques for the royal pleasure. How

[1] We hear of him as Master of the Singing-Boys at Westminster Abbey from 1480, as a Gentleman of the Chapel from about 1492, and from 1509 as Master of the Children of the Chapel.

far he was dramatist as well as producer is not clear. All his work for the stage has perished. Today he is known only as a composer who set to music verses both pious and Skeltonic, and in the absence of evidence the claim that he is an important figure in the history of our drama is excessive. Nevertheless, if John Rastell[1] and his son-in-law John Heywood are our first known lay dramatists, that is rather a token of our ignorance and a witness to the depredations of time. What is more important is that with John Rastell we feel ourselves to have arrived at a New Age.

In the first place, we must gratefully acknowledge his services and his son William's to our early drama. But for their presses both plays of Medwall, four of Heywood's, possibly *Magnificence*, and John's own, might have perished without trace or at best have joined the limbo of those which have survived only in title. His belief in the value of printed drama appears the more remarkable if we consider that of the eighteen dramatic pieces which got into print before 1534, no less than twelve were printed or published by him or his son. Fragmentary as is our knowledge of early sixteenth-century drama, without the Rastells it would have dwindled to a point. Moreover, John was not satisfied merely to print and to write plays. He built in the garden of his property in Finsbury Fields a stage of lath and timber and provided a wardrobe of players' garments, curtains, and linen cloths which were lent out during his absence in France, much to their detriment. His interest in the drama was of course didactic. In this medium he might serve the state by disseminating knowledge among a public that could not or would not read.

Secondly, Rastell was a patriot. The steady growth of

[1] John Rastell (1475?–1536) was bred in Coventry, acquired his knowledge of the law at the Middle Temple, and by 1504 had married Elizabeth More, sister of Thomas. Coroner of Coventry (1506), overseer of the transport of artillery in London (1514), shareholder in a voyage to the New-found Lands which owing to mutinous mariners got no further than Ireland (1516–17), garnisher of the roof of the Banqueting Hall at the Field of the Cloth of Gold (1520), trench-maker in France (1523), deviser of royal and civic pageants (1521, 1522, and 1527), chancery lawyer, and member of parliament (1529), he was a man of restless versatility. In his last years he became a henchman of Thomas Cromwell and an ardent advocate of reform in the common law and the Court of Chancery as well as in the Church; and perhaps because he exceeded in zeal and fell short in discretion—for Henry had no intention of making his new *Ecclesia Anglicana* a Lutheran church—he died in prison and in poverty. His probable dates as a printer are 1512–30: his son William printed from 1530 till 1534.

national consciousness, C. L. Kingsford maintained, is the real quality of fifteenth-century England, yet there is little evidence of it in earlier moralities. A sense of superiority is a common, though not a necessary or laudable, mark of patriotism; but these early dramatists hardly seem to be aware of the existence of foreign countries except for a rare reference to lechery as 'sumtyme gyse of Frawnce' or the satirical use of a French phrase or greeting often put into the mouth of a foppish Vice. But Rastell's patriotism is proud and positive. In *The Nature of the Four Elements*, an interlude which he wrote between 1517 and 1527, he expresses his belief that as the Greeks and Romans used their mother tongues for the dissemination of knowledge, so should we, 'our tongue maternall' now being sufficient 'To expoun any hard sentence evydent'. What is more, in the same play he confesses his belief in the manifest destiny of England to be a colonial power in the rich land of America, the land which Henry VII 'Causyd furst for to be founde', the land in which the shortsighted Henry VIII—Rastell seems to say by implication—takes no interest. What an honourable thing for king and realm if Englishmen had first taken possession and the king's dominion had been so far extended! And what a meritorious deed to instruct the natives in the true religion! But now French fishermen and others have found the trade and send a hundred sail a year. Rastell writes with the bitterness and longing of one who had himself attempted to cross the Western Ocean 'not longe ago' and had been thwarted by mutinous mariners.

Lastly, the emphasis of his play is on this world. As in Medwall's *Nature*, the first speech in *The Four Elements* is given to Nature, 'nature naturate' as distinguished from God who is 'nature naturynge'. The concern of both men is with the conflict between Good and Evil for the soul of man, both attach great importance to Reason, but whereas Medwall follows orthodox church doctrine, Rastell seeks the salvation of Humanitas by way of natural reason and human learning. To this end Studious Desire and Experience instruct him in astronomy, cosmography, and natural phenomena, the vices Sensual Appetite and Ignorance being inserted for the sake of those who are more disposed to mirth than to gravity. Rastell was shocked that men should know nothing of the universe in which they live. How presumptuous to call oneself a clerk and dispute on things invisible yet be ignorant of things visible! For this neo-Platonic rationalist the lowest step on the ladder which leads to God is a knowledge of God's creatures, of natural philosophy. In his belief the way to right action was an appeal to the reason. Characteristically, he refused to believe in the Trojan ancestry of Britain. As characteristically, he sought to prove the existence of Purgatory by what his opponent, the martyr John Frith, described as 'a wyt sophisticall whych he called naturall reason'. Perhaps it is also characteristic of Rastell that instead of his converting Frith, Frith converted him:[1] so providing the world with the rare spectacle of a change of heart in a man engaged in public controversy. In short, Rastell was a forward-looking man, whatever shade of approbation or disapprobation we may give to that epithet.

*The Four Elements* is by a quarter of a century the earliest printed attempt to teach astronomy in the vernacular. It is also the first English work to name America and the first (and possibly the last) attempt to give in dramatic form elementary lessons in science. And it provides the first English example of printed music—the music for the song and dance 'Time to pas with goodly sport'. But it is as far from being dramatic as any ill-versified dialogue can be. If Rastell were the author of two other interludes which have been ascribed to him, we should have to rank him higher. *Gentleness and Nobility* is a debate between a Knight, a Merchant, and a Ploughman, on a theme already handled in *Fulgens and Lucrece*. The play itself is written in four-beat couplets, unusually well controlled; the epilogue spoken by a Philosopher is in rhyme royal, a stanza made for state and stateliness. There is no action, unless the whipping which the Ploughman administers to his adversaries (*Et verberat eos*) can be called action. The arguments themselves are for the most part commonplaces of the age and of the later Middle Ages—that all are born to labour 'As a byrde is to fle', that what makes a gentleman is not birth, not inheritance, but 'vertew & gentyll condycyons'; but the debate is spirited, and each party to it, especially the Knight and the Merchant, receives many a shrewd knock. The Ploughman belongs to that vigorous tradition, descending from Langland, which under cover of a type criticizes fiercely and radically the existing

[1] A feat only exceeded in his century by the brothers William and John Reynolds who succeeded in converting each other (Pierce, *Marprelate Tracts*, p. 56).

structure of society. Was Rastell capable of such lively work? Or did it come from the more skilful pen of John Heywood? These are matters of debate. The striking suggestion that 'governours' and judges should be appointed for limited periods and then examined for partiality, found also in Rastell's *Pastime of Pleasure*, comes in the Philosopher's speech, and perhaps we may more certainly detect his hand in the epilogue than in the play itself.

To another play from his press he has been suspected of adding a conclusion. *Calisto and Melibea* is adapted from the Spanish. Any adaptation from the Spanish at this date, dramatic or non-dramatic, is remarkable. Still more remarkable is the original work, as remarkable if we consider it out of time and space as if we consider it in its setting in late fifteenth-century Spain. An extended prose 'novel' written entirely in dialogue form and divided into twenty-one 'acts', dramatic without being a drama, *La Celestina* is a work of genius which presents a story of romantic love against a background of grimly realistic comedy. The infatuation of Calisto and Melibea and their tragic end are the one theme: the shifts and devices of the subtle bawd Celestina and of Calisto's servants Parmeno and Sempronio are the other. This accomplished work an unknown Englishman chose to adapt. He took his material mainly from the first, second and fourth 'Acts', and 800 out of 1,088 lines, it has been calculated, come from the Spanish. He cut a large thong of another man's leather. We may note by the way that several Spanish proverbs—'Tomorrow is a new day', 'As soon goes the lambskin to market as the sheep's' (*Tan presto se va el cordero como el carnero*), 'The half knows what the whole means'—make their first known appearance in an English dress. So a language is enriched. It is more to the point to observe another expression that makes its earliest appearance, Calisto's line 'Yet worship I the ground that thou gost on.' Although in the novel (*adoro la tierra que huellas*), and in the play, the words are addressed to the bawd Celestina, they yet serve to indicate the ecstasy of passion represented by a character that is neither typical nor allegorical. *Fulgens and Lucrece* is sometimes called a play of romantic interest, but the play is of marriage rather than of love, and romance cools in the calm atmosphere of debate. Something of the passion of the original spills over into *Calisto and Melibea*. The infatuated

Calisto counts the world and heaven well lost: he is the first *jeune premier* of our drama.[1] But at the very moment when Melibea's will is weakening, the Englishman throws into this subtle work a moral spanner which brings the interlude to a jarring conclusion. A 'happy' ending makes the play a *comedia* more truly than the novel; it is described as a comedy on the title-page, the first printed play in English to be so called, except perhaps for the translation of Terence's *Andria* printed about the same year 1530. Melibea's father Danio (in the original Pleberio) has a terrible vision which brings Melibea to repentance, and the moral is pointed that parents should teach their children to say their morning prayers, and should preserve them from idleness by causing them to earn their living in some 'art craft or lernyng', and that laws should aim at removing the causes of crime, not merely at punishing the effects. The conclusion may be Rastell's, but it is in the same metre; this is the only play of its period consistently written in tetrameter rhyme royal.

In turning from John Rastell to his son-in-law[2] we turn from a man whose didactic purpose was only too obvious to one who seems at first sight to have no didactic purpose at all, from a man who is still writing in the allegorical framework to a man who has almost entirely rid himself of allegory. Thomas Warton, while he refused to concede to Heywood the merit of being our first writer of comedies or of being a writer of comedies at all, allowed that 'he is among the first of our dramatists who drove the Bible from the stage, and introduced representations of

[1] Gayley makes the further point that Melibea is the first secular woman introduced into drama, not as an object of scurrility and ridicule, but as dramatic material equal to that of men.

[2] John Heywood (1497–1579?) married the daughter of John Rastell. If Wood is right, he spent some time at Broadgates Hall, Oxford, but not long, 'the crabbedness of logic not suiting with his airy genie'. He was actively engaged at Court from 1519 to 1528 as a singer and player of the virginals. A pensioner of the Crown from 1528, he held minor office under the City of London, of which he became a freeman in 1523. His extant plays, four of which were printed by William Rastell in 1533–4, were probably written between 1519 and 1528. Under Edward VI and Mary he devised pageants and wrote plays in association with John Redford and Sebastian Westcott, some of them for performance by the children of St. Paul's, and also assisted George Ferrers and William Baldwin in Court plays. Convicted of treason in 1544 as a conspirator against the reformer Cranmer, he recanted and was pardoned. A good Catholic, as became a young friend of Sir Thomas More, he went into exile in 1564, entrusting his property to the care of his son-in-law John Donne, the father of the poet. He was still alive, and living at Mechelen (Malines), in 1578.

familiar life and popular manners'. The canon of his work is to some extent uncertain. We know on excellent evidence—the evidence of his brother-in-law William Rastell—that he wrote *The Play of Love* and *The Play of the Weather*; and a play without title which has been variously called *Wit and Folly* and *Witty and Witless* (and might also be called, from its three characters James, John, and Jerome, *The Three JJ*) exists in a contemporary manuscript, not autograph, which bears his name. Both *Witty and Witless* and *Love* are little more than debates. In the former the argument turns on the question whether a witless man is happier than a witty. James is a better dialectician than John, and not until the intervention of Jerome is the position established that the witty man, especially if he is wise, is preferred to the witless, both on earth and in heaven. *Love* is also an exercise in chop-logic and almost as devoid of dramatic action. Here there is a double debate: whether Lover-not-loved suffers more pain than Loved-not-loving (a woman) and whether Lover-loved is happier than No-lover-nor-loved. Each pair submits to the arbitration of the other, and the judgement is that there is nothing to choose between the pains of the one pair or the pleasures of the other. Thanks to the presence of 'the Vyse'[1] No-lover-nor-loved, there is more evidence in this play of the Heywood who claimed to have written 'many mad plays' and to have made Queen Mary smile. The Vice's trick of persuading Lover-and-loved that his mistress has been burnt to death leads to the famous direction: 'Here the vyse cometh in ronnyng sodenly . . . among the audyens with a hye copyn tank[2] on his hed full of squybs fyred cryeng water water, fyre fyre . . . tyll the fyre in the squybs be spent.' *Weather* is also disputatious, but some of the speeches are admirably in character. A quarrel among the gods who dispense the weather leads Jupiter to visit the world and enquire into the wishes of men and women. A stream of plaintiffs plead for the weather that suits their pleasures or professions: a gentleman, a merchant, a ranger, a water-miller, a wind-miller, a gentlewoman, a launder (laundress), and a boy, 'the lest[3] that can play'. The gentleman, as befits a gentleman, addresses the god himself, but the others direct their complaints to 'Mery reporte the vyse' whom Jove engages as his crier and go-between. All the characters drop in

[1] On this term see below, pp. 59 ff.

[2] *Copyn tank*, a high-cornered hat shaped like a sugar-loaf. [3] *lest*, smallest.

one by one except the two millers and the two women, and the clashes between these add variety to the 'merry' speeches of the Vice. At the end the weather remains as it was.

All three plays share certain characteristics. They conform to the usual length of the Tudor interlude—a thousand lines or a little more or less. All are debates, and *Witty and Witless* and *Love* in particular could appeal only to an audience that delighted in the conduct of an argument on a paradoxical theme. All end on a didactic and indeed a pious note, so that it is not true to say that Heywood aimed merely at satire and amusement. In all except possibly *Love* there is a tedious recapitulation of the arguments before judgement is given. In all the humour is boisterous and sometimes obscene. Those words of double meaning which George Puttenham advised a gentleman never to use in mixed company this dramatist delights in: in him we meet them in any quantity for the first time, but by no means for the last. In all, the dominant verse is the light four-beat couplet with occasional use of rhyme royal for dignity. One tiresome rhetorical trick—the love for which lasted till the end of the century—now makes an appearance in drama. The figure in which the changes are rung upon one word through several lines is perhaps more tolerable in comedy where the actor may gabble his tongue-twisters as fast as he pleases. In prose, so the barber said, it helped to unsettle Don Quixote's wits.

All or almost all these characteristics are to be found in *The Four PP*,[1] a play printed as Heywood's by William Middleton about the year 1544. This too is a debate. A Palmer, a Pardoner, and a Poticary dispute for precedence, and praise their way of life or display the relics and medicines they have to sell. A Pedlar who packs up his wares and takes holiday when he realizes he can sell nothing to this sophisticated company, advises the trio to join forces under the leadership of one, the leadership to go to him who can tell the greatest lie. By common consent the award goes to the unaggressive Palmer who protests that in all his travels he has never seen or known of any woman out of patience. But the Palmer refuses to lead, the Pardoner and Poticary mend their ways, and the play ends in a burst of orthodox piety.

For Heywood's authorship of *The Pardoner and the Friar* and

[1] *PP*, i.e. Ps, the duplication of the letter indicating a plural.

*Johan Johan*, so much more dramatic than the four plays we have considered and the best farces that have survived from early Tudor times, there is no contemporary evidence. They were first attributed to him in 1671 by that excellent collector and seller of plays and romances, Francis Kirkman, who recommended Heywood to his readers as 'the first English Playwriter' and as one who 'makes notable work with the then Clergy'. Yet there is little doubt that they are his. The resemblance between the two pardoners in *The Pardoner and the Friar* and *The Four PP* is close, some of their lines are identical, and a debt to French drama is by no means confined to *The Pardoner and the Friar* and *Johan Johan*. The surprising thing is not that Heywood showed a knowledge of French farce but that his should be the only surviving plays of his time that show this knowledge. There was much intercourse between the two nations, and in January 1529 the French ambassador reported as evidence of Wolsey's good will that the Cardinal had caused several farces to be acted *in French*. In this rich if sometimes rank pasture there was food for a score of dramatists, but only Heywood drew nourishment from it. The probability is that he knew not only some of the farces in which France excelled from the thirteenth century to Heywood's own day and which have survived in such remarkable profusion, but that he was acquainted also with the dramatic dialogues of the French stage. Heywood's partiality for the French drama accounts for his break with the morality convention. Characters that are not abstractions teaching the way to salvation but representations of ordinary men and women engaged in their lawful or unlawful vocations, though placed in the extravagant situations appropriate to farce; debates in which the pains and pleasures of love are weighed in the balance; farces in which a variety of men and women appear before a mythological personage; the tongue-twisting repetition of the same word; the extensive use of *double entendre*—these are some of the affinities which have been observed between Heywood and French drama. If over four of his plays he did not shed that 'total deprivation of moral light' characteristic of French farce, the reason may be that he was English. Not easily does an Englishman, especially a Tudor, resign himself to the sole purpose of amusing. A few years later in the century Des Périers recommended his *Nouvelles* to the reader with the words: 'Ventre d'ung petit poysson, rions! . . . riez seulement'; the English translator, who may have been Thomas Deloney, substituted a moral.

Of all Heywood's plays, the most insouciant—if that goes for an English word—are *The Pardoner and the Friar* and *Johan Johan*, and it is here that his debt to the French is greatest. *The Pardoner and the Friar* owes much to Chaucer, himself an admirer of the French spirit, but it uses one of the stock situations of French farce. A pardoner and a friar appear in or before a village church and preach begging sermons to the audience. The sermons are delivered simultaneously, with occasional breaks when the noise becomes intolerable and they turn to abuse each other. The same situation is similarly developed in a French farce—which may also have given a hint to *The Four PP*—where the two characters are a pardoner and a 'triacleur' or travelling apothecary. There is no verbal borrowing, yet the similarity is striking. But the two farces proceed to different conclusions. In the *Farce d'un Pardonneur, d'un Triacleur et d'une Tavernière* the two hypocrites make their peace at a tavern and cheat the hostess at departure: in the English, a curate and a constable, neighbour Prat, indignant at this pollution of their church and parish, attempt to drive them away. We prepare ourselves for the moral wrench which Heywood had given to his other plays, but it does not come. The constable with a bloody coxcomb and the parson with 'more tow on my dystaffe than I can well spyn' are pleased to let the intruders go 'with a myschefe'.

As devoid of moral instruction is *Johan Johan*. The three characters in this short interlude (678 lines) are Johan Johan the cuckolded husband, his wife Tib, and her lover, Sir John the priest. The wretched husband, worried by the absence of his wife, promises her a beating, but when she returns his courage evaporates, and he submissively fetches the detested and suspected Sir John to supper; and while Tib and her lover consume pie and wine he sits by the fire chafing hard wax for the purpose of mending a hole in a pail. His muttered comments which he dare not allow his wife to overhear and the remarks of Tib and Sir John as they enjoy the supper, remarks the double meanings of which are ill concealed, are superb farce. There is nothing which is not dramatic, nothing which does not illuminate the characters and the situation, and while the farce cries out to be acted it has a brilliance of dialogue which gives it a life over and above its life upon the stage. The light four-beat couplet, Heywood's favourite metre to which he adhered more faithfully than any other Tudor dramatist, is always sprightly. An acute critic has observed that the very title shows the dependence of this play upon French farce, for Johan Johan is but Jehan Jhenin, the diminutive in French signifying a sot or cuckold: the double meaning of 'chauffer la cire' ('to attend long for a promised good turne', Cotgrave) was also lost upon an English audience. Not until 1949, however, did the farce upon which Heywood depended come to light: *Farce nouvelle et fort joyeuse du Pasté et est à trois personnages. C'est à savoir: l'homme, la Femme, le Curé.* Of this farce *Johan Johan* is a translation—a brilliant translation, for Heywood thoroughly adapted it to English manners, with English allusions and English idioms and some of those familiar proverbs which he loved so well that he preserved as many of them as he could in a versified dialogue printed in 1546. At the end he takes his own way. In both farces the exasperated husband at last rebels and there is a scuffle, but only in Heywood does he drive the lovers off the stage. For a moment he enjoys his triumph; then the play ends as it began with Johan suffering the pangs of jealous suspicion.

Heywood was not the father of English comedy. Not from this soil did Shakespearian comedy spring. If others had followed his example, we might have had a school of English farce; but he had no disciples. What plays he wrote after the fifteen-twenties we do not know. He continued to entertain the court and he wrote or produced plays which children acted before Mary and Elizabeth. But William Rastell gave up his press in 1534, and all Heywood's later interludes are lost. We are left with this reflection, a 'cooling card' for English pride: the best of our early farces was translated from the French, and the best of our brief moralities (*Everyman*) from the Dutch; and if we wish to find the best of the diffuse moralities, we must go to Scotland. Not for more than half a century after *Johan Johan* do the plays begin to be written which entitle us to lift up our heads and our hearts.

### 4. *Papist and Protestant*

If we are surprised that we have to wait so long before the drama comes into its own, we are indulging ourselves in a belief in the necessity of progress. When we remember upon how many factors good drama depends—poets who are encouraged by the state of the theatre and the language to write dramatic poetry that will hold the stage, good actors, a receptive audience with a wide and liberal view of what constitutes entertainment, generous and reasonably intelligent patrons—we ought to be surprised not that the Great Age should have come late but that it should have come at all. The early years of Henry VIII's reign, when the monarch was playing Magnificence and Heywood was writing his interludes, appear at first sight to be propitious for drama and poetry as well as music. Lyric poetry and music do indeed flourish, for their themes are uncontroversial and timeless, and not even the wars and dissensions of the fifteenth century had been able to stifle them. But a public art like that of the drama is affected at every turn by the events of the time. To blame the theological and political controversies of the last twenty years of Henry's reign and of the reigns of Edward and Mary for the 'arrest' of English drama would be to assume that if England had remained Catholic a Marlowe or a Shakespeare might have appeared earlier. But what is certain is that in consequence of these controversies the drama became for a time not merely didactic but polemical. When Church doctrines became matters of dispute, plays turned to argument and invective.

For the propagation of theological and political doctrines, the morality, the sermon, and the ballad were the three obvious vehicles of popular instruction. We have seen that Skelton had used the morality play for political satire, and another play which has been suspected of satirizing Wolsey is *Godly Queen Hester*, where the plot came ready made to the dramatist from the Bible, and the evil councillor Haman tops in evil the vices of Pride, Adulation, and Ambition. The play would then belong to the late fifteen-twenties. We learn from Hall that a morality by John Roo acted at Gray's Inn at Christmas 1526/7 represented Lord Governance as ruled by Dissipation and Negligence, and displeased Wolsey; and we learn from John Foxe that Simon Fish, the author of *The Supplication of Beggars*, acted in an interlude which offended Wolsey, fled abroad, and so became associated with Tyndale. In May 1537 Cromwell was told of a play at a May game at Hoxston in Suffolk, about how a king should rule his realm, in which he who played

Husbandry said many more things against gentlemen than were set down for him. This was 'miching mallecho'. Innocuous but difficult to date is *Wealth and Health*. Entered in 1557 and surviving in an edition of *c.* 1565 with a prayer for Queen Elizabeth, it may be Henrician or Marian. The doctrine that Wealth, Health, and Liberty are national blessings, if unperverted by Ill Will and Shrewd Wit and guided by Good Remedy (a nobleman and a patriot), would not come amiss at any time in the century; nor would the sharp view that there was an excess of 'allauntes' (aliens) in the realm and that Englishmen would live the better without them.

The impact of German Protestantism was being felt in England by the early fifteen-twenties. The famous Bull of Leo X ordering the destruction of Luther's books was issued in 1520, and in the following year they were burned at Paul's Cross by the order of Wolsey. The authorities were now watchful for the importation of heretical books, but so long as England was orthodox, anti-Protestant drama was unnecessary and might even have served to air doctrines better suppressed. It is true that in 1527 the boys of St. Paul's School under the High Master John Ritwise acted at Court before ambassadors from France a play in which Wolsey was glorified and Luther and Luther's wife were ridiculed, but it was in Latin and French and in no sense popular. The situation was soon to change. Wolsey fell in 1530, between that year and 1534 Henry broke with Rome and became in name and in fact the Supreme Head of the Church, and by 1539 the monasteries had been suppressed. For a time, as he flirted with the German and English Protestants, it appeared that he was to throw over Catholicism as well as Roman Catholicism, and Thomas Cromwell during his chancellorship (1532–40) went much further than his master warranted in his encouragement of 'the new learning' of Protestantism. He had nothing to learn in the crafts of propaganda, and under his protection the indomitable, indefatigable scholar John Bale wrote the bitter interludes which display violent enmity towards the monstrous regiment of Papists.[1]

[1] Born in Suffolk of poor parents and sent at an early age to the Carmelite house at Norwich, Bale (1495–1563) was educated at Jesus College, Cambridge, where he met Cranmer, took the degree of B.D. in 1528–9, and before taking that of D.D. (*c.* 1533) travelled in France and elsewhere. He abandoned his monastic vow in the early thirties, became a parish priest and a Protestant, and

The reader of Bale's interludes has difficulty in believing that they were acted, yet many were eager to hear the new doctrine, and in the days when sermons lasted at least an hour and a preacher did not scruple to turn his glass and begin on a second hour, an audience might well be glad to listen to the same doctrine in dramatic form. While still a parish priest Bale was writing anti-Romanist plays in English in the service of John Vere, Earl of Oxford. But his years of greatest dramatic activity, years in which he revised old plays, wrote new ones, came under the patronage of Cromwell, and organized a company of players, were 1538 and 1539.

Five of his twenty-one plays have survived, four in print. Of these, three form a trilogy, *God's Promises*, *John the Baptist*, and *The Temptation of our Lord*, and it is remarkable that on the day of Mary Tudor's coronation these plays were acted at the Market Cross at Kilkenny, the first and longest in the morning and the others in the afternoon, 'to the small contentacion of the prestes and other papistes there'. In these 'plays' there is little but straight doctrine and their genre is in part that of the miracle play, *God's Promises* being a kind of cycle in seven parts, and in part that of the interlude employed in the service of religion. *The Three Laws* is more ambitious and, in the mocking satirical manner of the late morality plays, characters like Infidelitas, Sodomy, Idolatria, Avaritia, Ambitio, Pseudo-doctrina, and Hypocrisis provide a ghastly mirth. It may claim—if the claim is worth making—to be the first *Protestant* morality play. The end of the play was revised in the reign of Edward VI, but if the rest of it was compiled, as the title-page asserts, in 1538, it may also claim to be the first English play to be divided into five acts. Here again the claim is hardly worth making, for an act to Bale was but a part, and it was nothing more. Act I is introductory, the laws of Nature, Moses, and Christ give him Acts II–IV, and Act V shows the punishment of Infidelitas by Vindicta Dei, the banishment of Babylonical Popery, and the

married. Imprisoned for his doctrine in 1536 and released through the intercession of Cromwell, he was ejected from his living at Thorndon in Suffolk in 1537. On Cromwell's fall he took refuge in Lower Germany, but returned to a Protestant England in 1548, and (against his will) was appointed in 1552 to the see of Ossory. Thence he was forced to flee, after a few months, upon the accession of Mary, and after many adventures contrived to reach the Netherlands. He returned to England, but not to Ireland, in 1559, and as a canon and prebendary of Canterbury ended his turbulent life in peace and in the search for England's antiquities.

triumph of Fides Christiana. In Bale, as in so many other controversialists of his time, Protestantism went hand in hand with loyalty to the crown and an ardent nationalism. It is a little ironical that during his second exile this strident Anglican was shocked by a group of extreme Calvinists who regarded the English Communion as a Popish Mass and felt no special allegiance to country and to crown.

The best known of his plays by reason of its resemblance to the Elizabethan history plays is *King John*. It survives in a manuscript revised, and for the last thousand lines rewritten, by the author, a revision made after 1558 perhaps for a performance before the Queen at Ipswich in 1561. (If she saw it, we may wonder whether she cared for its doctrine and whether she remembered her proclamation of 16 May 1559 that no writers of interludes were to touch on matters of religion or of civil government.) The play may not differ in essentials from that performed at Cranmer's house in 1539 when at least one member of the audience was convinced by Bale's polemics that John was 'as noble a prince as ever was in England'. Even in 1539 it may not have been new, for already by 1536, as he tells us in his *Anglorum Heliades*, he had written a play *Pro Rege Joanne*. Some have observed a resemblance between *King John* and *The Three Estates*, but apart from the introduction into both plays of the three estates there is no resemblance. There is a world of difference between the humane and humorous genius of Lindsay, with his wide sweep and genuine sympathy for the suffering poor, and the bitter doctrinaire spirit of Bale. We cannot imagine Bale sparing even a reformed Wantonness, Placebo, and Solace.[1] Tudor playwrights might well have profited from Lindsay's play, if they had known it, but as we have seen it did not get into print till 1602, and then it was too late. The bitter *Pammachius*, which the German Protestant Thomas Kirchmeyer (Naogeorgus) dedicated to Archbishop Cranmer, Bale did know, for his translation is one of his many lost works. Kirchmeyer was a cock of the same hackle, and his Latin play, in which the Pope is identified with Antichrist and contemporary history manipulated to serve allegorical needs, yields nothing to Bale in the bitterness of its polemics. (It was performed at Christ's College in 1545—at a cost of twenty nobles—during the

[1] See above, p. 18.

vice-chancellorship of Matthew Parker, and was the cause of a breach between him and Bishop Gardiner, then Chancellor, which was never healed.) But *Pammachius* belongs to the year 1538, by which year the bent and bias of Bale's work were set. From one English work he may well have taken a hint, Tyndale's *Obedience of a Christian Man* (1528). Not only did Tyndale hold, like Bale and like St. Paul (Rom. xiii. 1–2), that a king was God's representative, whom none might judge but God Himself, but he regarded John sympathetically as the victim of Popish historians. Two years later the author of *A Proper Dialogue between a Gentleman and a Husbandman* was maintaining that John was a Protestant hero who defended the true religion against the assaults of a corrupt clergy.

Some modern historians maintain that John was not so black as he was painted, but no one since Bale has painted him pure white. John becomes an idealized Christian hero, the noble champion of an England widowed of her husband, God, by false religion. Nobility and Civil Order (the Law) are at first faithful, but Sedition and Dissimulation, Private Wealth and Usurped Power, seduce them from their allegiance; with the aid of the Clergy John is subdued, and England has to wait three hundred years for a 'Duke Josue' to lead her to the land of milk and honey. In the second act—for the play in its revised state is in two acts or parts—John is poisoned by Dissimulation, and the simplest lines in the play come in his farewell speech to England:

> Farwell swete Englande, now last of all to the
> I am ryght sorye, I coulde do for the nomore
> farwele ones agayne, yea, farwell for evermore.

But the play ends on a cheerful note. Verity defends the character of John from the slurs of Popish chroniclers, especially Polydore Vergil with his Romish lies; Nobility, Clergy, and Civil Order repent; Imperial Majesty exiles Private Wealth from the monasteries and Usurped Power is allowed to 'goo a birdynge for flyes'; but stubborn Sedition, who will not abandon the Pope, is hanged and quartered. At times these abstractions assume the names of historical figures. Sedition becomes Stephen Langton, Private Wealth Pandulphus, Dissimulation the poisoner Simon of Swinstead, and Imperial Majesty will stand equally well for Henry, Edward, or Elizabeth. Bale's originality rests in his choice of a theme from English history as the base for a political morality. In a sense, but not perhaps an important sense, he is a precursor of the historical plays on English themes of Elizabeth's last years. His play remained in manuscript, however, and as far as we can tell had not the slightest influence upon later drama. A contemporary dramatist went to the Bible in his *Godly Queen Hester* for a heroic queen and a seditious counsellor: Bale went to English history. He went to English history for another play, the lost *De Imposturis Thomas Becketi.* A Canterbury pageant on this theme had been revived in 1504–5, acted by children at various points in the city, and acted with some realism, for two bags of red feathers represented the blood of the martyr. In the pageant St. Thomas was a saint: in Bale the villain of the piece. But for his purposes a heroic John was far better than a heroic Hester or a wicked Thomas, and he took every chance of representing through John the fight of Protestant and patriotic piety against the encroachments of Rome. Every chance, that is, except one; a diatribe is not a drama, and Bale's attention is always and wholly concentrated on his doctrine. His verse never rises above doggerel, and even in those parts of *King John* which were written after 1558 we should be lost were it not for the strong caesura which splits each line into two mechanical halves. Perhaps it is not so very unfair to take as characteristic of his doctrine, his tone, and his metre two lines from *Three Laws*:

> In no case folowe, the wayes of Reygnolde Pole,
> To hys dampnacyon, he doubtles playeth the fole.

The polemical play flourished during the ascendancy of Cromwell, in the reign of Edward VI, and during the earliest years of Elizabeth's reign; but apart from Bale the remains are scanty. Whether our loss is serious may be guessed, perhaps, by what we know of the work of Ralph Radcliff. From 1538 he conducted a private school at Hitchin in a building which had once been the property of the Carmelite friars. When the ex-Carmelite John Bale visited the school, he professed himself pleased: *mihi . . . multa arriserunt: eaque etiam laude dignissima.* On the stage which Radcliff provided, his boys acted *De Jo. Hussi damnatione, De Sodomae incendio,* and plays on other biblical themes compiled by their master, whether in English or Latin we do not know. A closer friend of Bale's and a fellow-exile,

John Foxe, came under the influence of the thriving school of Latin drama in Germany and wrote a play not *more veteris comediae* but after the manner of Kirchmayer. In his *Christus Triumphans* (Basel, 1556) the papal Antichrist Pseudodamnus corresponds to Kirchmeyer's Pammachius, and a courtesan Pornapolis (Rome) is the *meretrix* of the Apocalypse. His play ends with an Epithalamium celebrating the marriage of Christ and Ecclesia. It was acted at Trinity College, Cambridge, in 1562–3.

In English, however, we have very little of this nature after Bale. R. Wever's *Lusty Juventus* belongs to the reign of Edward VI. He would not care to know that we remember only the lyric

In a Herber grene, a sleepe where as I lay
The byrdes sang sweete in the myddes of the day

which gets into all the anthologies, and the 'gyrle nyse' who leads Juventus astray and whose real name is Abhominable Living. 'In youth is pleasure', Juventus is told, is a doctrine not warranted by Scripture, and Scripture, not the old traditions made by man, is man's only guide. Salvation lies in Faith. Though good works are the necessary fruits of repentance, they will not save a man (as they saved Everyman). Another play of the same reign of which only two leaves have survived—it has been given the title *Somebody and Others* or *The Spoiling of Lady Verity*—introduces a Papistical priest (Minister) and his friends Avarice and Simony. They oppress People and Verity, but some day Somebody will release People from these bonds and restore Verity to power. One play has survived[1] from the early years of Elizabeth's reign, *New Custom*. New Custom is another name for New Learning, the derisive term used by the Papists against the Protestants, but in the play New Custom observes that his age is 'a thousande and a halfe' and his true name Primitive Constitution. With the help of Light of the Gospel, Perverse Doctrine and his evil associates, Ignorance, Hypocrisy, and the two rufflers Cruelty and Avarice, are worsted, Perverse Doctrine is converted to Sincere Doctrine, and Edification, Assurance, and God's Felicity establish him in the faith.

[1] There are tantalizing references to the use of the drama as propaganda in the earliest years of Elizabeth's reign. In 1559 the Spanish ambassador complained to the Queen that London comedies were mocking his royal master and that William Cecil was supplying them with themes. See also Hawarde, *Camera Stellata* (ed. W. P. Baildon, p. 48) for Cecil's advocacy of drama for propagandist purposes.

Against all the Protestant polemical plays which we know of in title or in text we can set only one play written for the opposition, but as drama it is worth all that Bale and his successors wrote. *Respublica*, which has survived in one manuscript, belongs, as we should expect, to the reign of Queen Mary. In this 'Christmas devise' acted by children and designed to recreate the spirits of a noble audience—pretty certainly those of the Queen and her Court on New Year's Day 1554—the poet who speaks the Prologue announces his intention of showing—not 'by plaine storye But as yt were in figure by an allegorye'—how Respublica comes to ruin and decay when ruled by Avarice, Insolence, Oppression, and Adulation. These vices by cloaked collusion and under the counterfeit names of Policy, Authority, Reformation, and Honesty (respectively) flourish for a while until 'tyme bringeth truth to lyght'. Verity, 'the daughter of sage old Father Tyme' and her three sisters Mercy, Justice, and Peace, amend what is amiss and restore Respublica ('good Englande') and People ('the poore Commontie') from their late decay; and Nemesis, 'the goddes of redresse and correction', who is identified in the Prologue with the Queen, passes sentence upon the evil-doers. The burden of the play is Respublica's 'O lorde, howe have I bee used these five yeres past', and while Avarice and his three companions are typical and in no way particular, the audience must have been reminded of the calculating politicians of Edward's reign, who 'feathered their nests'—a favourite expression in the play—from the public purse. The play does not concern itself in the least with doctrine but mainly with the pockets of the people. The change of government (and incidentally of religion) brings the country from woe to wealth and reforms all the evils by which Respublica has been so notoriously abused:

> Usiree, perjuree, pitcheree, patcherie,
> Pilferie, briberee, snatcherie, catcherie,
> Flatterie, Robberie, clowterie, botcherie,
> Troumperye, harlotrie, myserie, tretcherie.

Poor People—an honest countryman who speaks in the conventional stage-dialect of a southwestern type—has much to complain of in the rise in prices, the encroachment of lands, the debasement of the coinage, de-afforestation so that poor folk can scarcely get a stick to make a fire, and 'great grazing'. Some of his complaints are very like those of Lindsay's Pauper in another *res publica,* but here the blame is not laid upon the Church: when priests and bishops had their livings, 'men were bothe fedde and cladde'. When Avarice and his gang are exposed, People is at once able to buy a new coat and to boast of a silver groat in his purse: 'I wis iche cowlde not zo zai these sixe yeares afore.'

To the historian of the drama the vital difference between *Respublica* and the Protestant plays is not that here the vices are Protestants and there they are Papists, but that here we have what is first and foremost a play, and there we have doctrine and invective expressed in dialogue. This author has a sense of shape and a sense of character. If not the first, his play is among the earliest English plays to be divided into acts and scenes on the classical model. The first act is to disclose the four vices, the second to show the deception of Respublica, the third to exhibit the vices in full swing and sway, the fourth to expose the sorry state of Respublica and People, and in the fifth the four virtues come to the rescue. A new scene is marked at the entrance of each character, and acts and scenes follow each other in a logical and dramatic sequence. The dramatist makes great use of what has been called the *lupus in fabula* formula ('And loe where Avarice comth a woulff in the tale'), a formula for which he could have found precedents in vernacular and continental drama, and above all in Terence and Plautus. For another peculiarity, which we find in this play for the first time in English, he could have found models in the texts of Terence that he read: the addition to the names of the characters of descriptive labels. Avarice is 'the vice of the plaie'—a cognomen which will be considered later[1]—Insolence is 'the chief galaunt', Oppression and Adulation are 'an other' and 'the third' gallant, Respublica is a widow, and the virtues are 'foure Ladies'. The labels suggest, what the text bears out, that the characters, especially the vices and People, are not abstractions but nicely differentiated types of men and women in contemporary society. This writer has a sense of comedy which is not suppressed but sharpened by his satirical purpose. His work is the nearest thing that England can offer in the morality tradition to *The Three Estates.*

[1] See below, pp. 59 ff.

To say that the play's value rests solely in its comedy would be unjust. The allegory which is invented is admirable for its purpose, and the comic characters are not allowed to get out of hand or to obscure the sorrows of Respublica and the sufferings of People. Occasionally we hear a higher strain, as when Avarice enquires of Justice what is her name (V. ix):

*Iust.* Iustice is my name.
*Avar.* Where is your dwelling?
*Iust.* In heaven and thens I came.
*Avar.* Dwell ye in Heaven, and so madde to come hither?
All our hucking[1] here, is howe we maie geate thither.
*Iust.* I bring heaven with me and make it where I am.

But it is the comic invention that makes the play. The author gives us that overplus or bonus of fun which we have no right to count on in satirical allegory, but welcome with open arms on the very rare occasions when it is given to us. The fun is not merely in the situation but in the words, and any actor lucky enough to play the part of Avarice will rub his hands at the passages where Avarice, in fear for his gold, complains of 'whoreson theves' and covetous knaves or the multitude of beggars crying 'geve, geve, geve, geve. Finde we oure money in the strete doo theye beeleve?'; or where his soul is half out of his body in the contemplation of his money, so that his friends can hardly make their presence known to him. Even the scene in which Avarice teaches the three 'galaunts' their aliases, elsewhere so dully contrived, even that scene becomes comic as Avarice grows more and more angry with his thick-headed pupil, Adulation. The promise of the Prologue that the actors 'shall make youe laughe well' was not broken. If the uproar in Christendom could have been stilled by laughter, plays like *Respublica* might have done it.

There is a natural reluctance to allow a work that stands above the ruck to remain anonymous. We do not like to see valuables lying about unattached. Was this play written by the man whose prologue to another children's play, probably of the same year, preached that 'Mirth is to be used both of more or lesse' and observed that Plautus and Terence 'Under merrie Comedies secretes did declare'? We know Nicholas Udall to be the author of *Roister Doister* only by a lucky reference in the

[1] *hucking*, bargaining, endeavour.

third edition of Thomas Wilson's *Rule of Reason*, also of 1553. If he wrote all the plays that have been attributed to him—*Thersites*, *Jack Juggler*, *Respublica*, *Jacob and Esau* as well as *Roister Doister*—he would become the most important figure in comedy between Heywood and Lyly, indeed between the Wakefield Master and Lyly. Three of these plays, together with *Roister Doister*, will be considered later.[1] What may be said here is that external evidence favours, if it does not confirm, his authorship of *Respublica* and that the internal evidence is good, but not decisive.

### 5. '*Morals Teaching Education*'

If only as an antidote to Bale and his like, this chapter will end with an account of John Redford's *Wit and Science*. To see the play in its proper setting we must know that its author—musician, composer, poet, and dramatist—was appointed master of the choir-school of St. Paul's between 1531 and 1534 and remained in that office until his death in 1547. Between these dates his play was written and performed. The interesting manuscript in which it survives contains poems by John Heywood, Sebastian Westcott (who was Redford's executor and residuary-legatee and succeeded Redford as Master), Thomas Prideaux, John Thorne (organist at York Minster), all men who had or may be suspected to have had a connexion with the choir-school of St. Paul's. Among the verses in this collection are some which Redford put into the mouths of his children—their lamentations at the cruelty of their Master and at their sufferings in learning 'this pevysh pryk song'.[2] The refrain at the end of each quatrain is 'We lytle poore boyes abyde much woe'. It is a pleasant song which could only have been written by a man who was both humorous and humane. One of his 'lytle poore boyes' at St. Paul's was Thomas Tusser. In his *Five Hundred Points of Good Husbandry* he praised Redford's skill as a musician and virtue as a man, and grieved that he passed out of Redford's care to Eton and the brutality of Nicholas Udall.

*Wit and Science* is one of the purest allegories that have come down to us. To the sixteenth century metaphor was allegory in a single word: the double meaning systematically sustained over a whole action or series of actions became allegory or

[1] See below, pp. 93–96; 105–9.

[2] *pryk song*, prick-song, the complicated descant or 'counterpoint' accompanying a simple theme or plain-song. Several pieces of this kind by Redford have survived.

expanded metaphor. In Redford's play not only is there a total absence of characters that are non-allegorical, but the invented action is so perfect in its representation of the meaning that the play becomes one long and continued metaphor. In other words the double meaning of the allegory is embedded in the action, and explanatory glosses are rendered superfluous. The play presents the wooing of Science (i.e. Knowledge), daughter of Reason and Experience, by Wit (i.e. Intellect, Understanding), son of Nature. The first task of the young and tractable Wit is to defeat the monster Tediousness, Science's greatest enemy. To assist him in the task, Reason (who favours the match) appoints as guides Instruction, Diligence, and Study. Wit in the pride of uninstructed youth disobeys Instruction, and attacks Tediousness with fatal consequences. Restored by Honest Recreation, Quickness, and Strength, all gifts of Reason, he fails once more and becomes a prey to Idleness. In her lap he falls asleep, as one who has *neque vox neque sensus*. The scene in which Idleness attempts to teach her daughter 'Ingnorans' the spelling of her name (Ing+no+ran+s s s) does not further the action but adds to the fun of the play. Meantime the sorrowful Science finds no comfort in Fame, Favour, Riches, and Worship, and when Wit appears before her and her parents in the garments of Ignorance he is cast off. But he catches sight of his image in the glass of Reason, is whipped by Shame, becomes truly penitent, submits himself to Instruction, Study, and Diligence, kills Tediousness near the foot of Parnassus with the sword of comfort, and is married amid much rejoicing to his Lady Science.

That the play can still please—at least in places where Wit and Science are valued—is proved by recent revivals on both sides of the Atlantic. It is brightened throughout by song and instrumental music. The verse is mostly in a fast-moving four-beat line, with short two-beat lines for Tediousness and for Wit's converse with Honest Recreation. It is never heavy and it is always speakable. As lively are the language and the characterization. Like Bunyan, Redford is sometimes so interested in the human situation that while never relinquishing his grip on the allegory he adds touches beyond the reach of allegory. So when Reason takes leave to doubt whether his daughter will accept the twice-fallen Wit, 'as women oft tymes wylbe hard hartyd'; so when, after troth has been plighted, the Lady fears her lover may find marriage a clog and is assured that 'to kepe you swetehart as shall be fyt Shalbe no care but most joy to wyt'. It is like Bunyan's Mercy at the House Beautiful explaining that she might have had husbands; but they did not like her conditions, 'though never did any of them find fault with my person'. In the play Wit is Wit and also a flighty young man who sobers to Learning, and Science is Knowledge and also a young damsel of great attractions.

Some twenty years after Redford's death two men were attracted by the ingenuity of his plot and attempted adaptations. They did not better his instruction. Their plays may be considered here, though they take us well beyond the limits of Henry's reign. One survives in manuscript—*The Contract of a Marriage between Wit and Wisdom* probably by one Francis Merbury; it is a vulgarization of Redford's play.

The other—*The Marriage of Wit and Science* (entered and printed in 1569 and probably the *Wit and Will* acted at Court at Christmas 1567–8)—is more respectable, mainly because it keeps more closely to Redford's admirable plot. It is divided into acts and scenes, but it is of about the same length (some 1,200 lines). The chief difference in plot is that in place of Confidence, who plays a very minor part as Wit's servant in the earlier play, Nature provides her son Wit (aged about 17) with an attendant named Will (almost our Wilfulness), aged between 11 and 12: he adds to the gaiety of the piece. The greatest difference of all is in metre and in rhetoric, matters which will be discussed in their place.[1] We contemplate the thought of children acting *Wit and Science* and *The Marriage*, then or now, with pleasure; but when we read the lines which Middleton and others wrote for children's companies in the early years of the seventeenth century we suffer many a twinge. As a heathen poet said, *Maxima debetur puero reverentia.*

There may have been a vogue for plays of this kind, and it is certain that more have been lost than have survived. By 1592, if we may believe Robert Greene in his famous and infamous *Groatsworth of Wit*, the vogue was over. There a player informs 'Roberto' that nowadays

> The people make no estimation,
> Of Morrals teaching education.

[1] See below, pp. 73 f.

But by that time the people had no liking for 'Morrals' of any kind. Their day with 'the people' was over. But in academic circles the taste survived. Thomas Tomkis's *Lingua: Or, The Combat of the Tongue and the Five Senses for Superiority* calls itself 'A pleasant Comedie', but it is a moral and it went through six editions between 1607 and 1657 and was acted with applause at school and university.[1] We could do with more moralities like *Wit and Science* and *Lingua* which teach 'severe Philosophy to smile'.

[1] Cf. the Cambridge *Pedantius* and the Oxford *Bellum Grammaticale*, both praised by Sir John Harrington in 1591.

# II

# THE LATE TUDOR MORALITY PLAY

## 1. *Authors and Actors*

PROFESSOR W. R. MACKENZIE has defined the morality as 'a play, allegorical in structure, which has for its main object the teaching of some lesson for the guidance of life, and in which the principal characters are personified abstractions or highly universalized types'. In a strict classification some of the plays mentioned in this chapter are not moralities but plays with morality elements: they have for their chief characters neither personified abstractions nor highly universalized types. Lewis Wager's *Mary Magdalene* is only to be distinguished by its inferiority from the Digby miracle play on the same subject (*c.* 1490). Both contain morality characters—for the Digby play is late enough to have been influenced by morality drama—but neither conforms to a strict classification. *King Darius*, which is inexcusably mean in plot and language, begins as a morality play and ends with the debate in (apocryphal) 1 Esdras iii–iv: the two plots meet in the middle but scarcely mingle. *Nice Wanton* is about a foolish mother and her three children, one good and two bad; the morality element is there in the character of Iniquity, and no play is more bent on teaching a lesson, but the base is non-allegorical. On the other hand, Nathaniel Woodes's *The Conflict of Conscience* seems at first sight to be even further removed from the strict morality form. The plot was suggested by the recantation and consequent despair of Francis Spira, an Italian lawyer, but Woodes suppresses the name from his play because he does not think it permissible in 'Comedy' to touch the vices of a private man but above all because he believes that his audience will be readier to find themselves in this character if the name is generalized from Spira to Philologus. And so, though the play is founded on historical fact, and introduces friends and children, a cardinal and an ignorant priest, it is substantially an orthodox morality play: for Philologus is a universalized type assailed by personified vices like Hypocrisy, Tyranny, Avarice, and Horror. But these are the refinements of modern scholarship and would not have occurred to a sixteenth-century audience.

Some plays admit a modicum of morality elements, but are distinguished by a new metre and a new rhetoric. The earliest of these belong to the fifteen-sixties. They are predominantly secular and non-allegorical, and derive their plots from legendary and historical themes: one (*Susanna*) goes for its theme to the Bible. Some of them are plays which must be considered in any account of the first fumblings after romantic comedy (*Common Conditions*, *Clyomon and Clamydes*) and after tragedy and tragicomedy (*Susanna*, *Patient Grissell*, *Cambyses*, *Appius and Virginia*, *Horestes*);[1] but they cannot be wholly excluded from this chapter and they enter especially into the sections on the Vice and on metre and rhetoric.

There can be few who read the morality plays or plays with morality elements which have survived from the reigns of Edward VI and Mary and the first twenty-five years of Elizabeth's reign. The plays of this period which are most read and have any chance of performance in our time are those which do not observe the allegorical convention and are to be considered later (e.g. *Roister Doister*, *Gammer Gurton's Needle*, *Jacob and Esau*). Some of the moralities of this date, as we have seen, are battlecries in the fight between Protestants and Papists,[2] but most of them are moral rather than controversial. It is true that nearly all are Protestant in tone. *Impatient Poverty*, which is certainly as early as Mary's reign and may go back to Henry's, is one of the very few that might be acceptable to both parties. But while most of them are Protestant, they differ from the polemical plays of Bale and others in that they make only incidental side-thrusts at the Papists. After the earliest years of Elizabeth's reign the dramatists seem to have been content to leave theological controversy to the Jewels and the Foxes of the age. They were encouraged to be non-polemical by Elizabeth's proclamation of 16 May 1559 that they were not to touch on matters of religion or of civil government.

The settings of these plays have been closely studied by Chambers and more recently by T. W. Craik. Almost all of them are indoor plays and many require no scenery at all, only a clear acting space or *platea*, often referred to as 'the

[1] See below, pp. 143–6. [2] See above, pp. 33 ff.

place'. The audience was too close to be ignored, even if dramatists had wished it, and as will be seen later it is frequently addressed from the stage, especially by the comic characters. Appropriate costume and symbolical colour made identification easy and added to the significance of the moral or the satire.

Most of these plays are anonymous, and where a name is attached—Ulpian Fulwell, Thomas Ingelend, Lewis and W. Wager, George Wapull, R. Wever, and so on—it is but a shadow of a name. Pretty clearly no man of name and fame cared to write moralities for the mid-sixteenth-century stage. In a book of this kind they must be given summary treatment. A few deserve more, but most are best considered—as James I dubbed Francis Bacon knight—'gregarious in a troop'.

Morality plays continue to be as short as they were (with one or two notable exceptions) under Henry. John Rastell, the only dramatist who was his own printer and publisher, mentions playing time and he alone in these first hundred years. His *Four Elements*, he tells us, may be played in an hour and a half, and if some of the 'sad mater' be omitted from the speeches of the Messenger, Nature, and Experience, 'yet the mater wyl depend conveniently' and the time be reduced to three quarters of an hour. Few Tudor moralities, early or late, run to much more than a thousand lines. But when from the early sixties a vogue sets in for secular plays with an admixture of morality elements—*Cambyses*, *Common Conditions*, and others—plays tend to run to two thousand lines and demand the 'two-hour traffic' of the stage, an innovation of much significance.

Some interesting changes may be observed in the kind of information which authors (or publishers) thought fit to give their readers. For some thirty-five years from 1547–8 plays advertise, usually on the title-page, the number of actors required and how the parts may be doubled, trebled, and even septupled. A note at the end of the Croxton *Play of the Sacrament*, a miracle play assigned to the late fifteenth century, states that 'IX. may play yt at ease', and the manuscript of Bale's *King John* contains frequent notes that an actor on leaving the stage is to dress for another character; but the first printed plays to give this kind of information are Bale's *Three Laws* (1538) which required five actors, *Impatient Poverty* (*c.* 1547–58), and *Wealth and Health* (1554), each of which states that four men 'may easily play this Play (Interlude)'.[1] The advertisement on the title of R. Wever's *Lusty Juventus* (*c.* 1565) is more laborious: 'Foure may play it easily . . . so that any one take of those partes that be not in place at once', an obvious proviso not always observed. Eighteen plays printed between 1547 and 1581 have survived that tell us the number of actors required; and of these, thirteen add the division of the parts. Seven demand four actors, three demand five, four six, one seven; and the three that require eight are late plays that have few morality characters—*Patient Grissell* (1569), *Cambyses* (1569), and Thomas Garter's *Susanna* (1578).

These plays were offered for anyone to act, and it is usually impossible to determine the kind of dramatic company—adult or juvenile, amateur or professional—from which they emanated. Some of them, indeed, emanated from no company. Nathaniel Woodes, minister in Norwich, wrote his *Conflict of Conscience* for performance 'in private houses or otherwise', and we may doubt whether any actors were tempted to put his intractable piece upon the stage. Some plays were performed by adult companies of professional players, if we may believe the evidence of *Sir Thomas More*. In this play of the fifteen-nineties the Cardinal's players (four men and a boy, the boy impersonating three women) offer the guests assembled for a banquet a choice between the following: *The Cradle of Security*,[2] *Impatient Poverty*, Heywood's *Four PP*, *Dives and Lazarus*, *Lusty Juventus*, *The Marriage of Wit and Wisdom*, *Hit the Nail on the Head*, all (except the last) extant or known from other references to have existed, and all except Heywood's probably belonging to the middle years of the century. But these plays, some of which may have originated as children's plays, may also have been popular with amateur and semi-professional actors, and the information as to the number of actors and the division of the parts is more likely to have been useful to them than to the real professionals. Strype in his edition of Stow's *Survey* (1720) observes how in these early days of the drama, before what 'was once a Recreation . . . by Abuse became a Trade and Calling',[3] ingenious tradesmen and gentlemen's servants in

[1] The dates are dates of publication.

[2] *Security*, ill-founded confidence, carelessness. For more evidence that this play was acted by professional players, see below, pp. 76 f.

[3] Also below, p. 52.

London 'would sometimes gather a Company of themselves, and learn Interludes, to expose Vice . . . And there they played at certain Festival Times, and in private Houses at Weddings, or other splendid Entertainments, for their own Profit'. In 1553 the City fathers forbade artificers and handicraftsmen to abandon their occupations and wander about singing 'Three Men's Songs' in taverns and at weddings and feasts to the detriment of the City minstrels: but they might continue to sing songs in common plays or entertainments. So in England today a butcher, a baker, or a candlestick-maker may spend his nights in a semi-professional jazz band blowing out his brains upon the saxophone. The recent discovery that the churchwardens of St. Botolph, London, let their parish hall to players between 1557 and 1568, during which years it may have been used a hundred times, is only the most striking of many scraps of evidence that point to the activities of amateur and semi-professional actors.

*Susanna* (1578) and *The Conflict of Conscience* (1581) are the last plays to announce the number of actors required and the division of the parts. Soon afterwards, on the threshold of the great age of the drama, when Elizabeth's reign was half over, the practice began of stating on the title-page the acting company and sometimes the place and date of performance. It is seemly that the publisher of *Gorboduc* (1565) should be the earliest so to do: and would that all titles had been so precise about the place (Whitehall), the date (18 January 1561–2) and the actors (Gentlemen of the Inner Temple)! Of non-academic plays that give information of this kind the earliest are Edwards's *Damon and Pythias* (1571), Lyly's *Sappho and Phao* and *Campaspe*, and Peele's *Arraignment of Paris*, the last three all published in 1584 as acted by the children of the Blackfriars or St. Paul's. (It is also seemly that these last three plays should be the earliest to be printed in fair roman type with nothing 'Gothic' about them. By the early nineties roman had supplanted black letter.) As for the plays acted by professional actors Robert Wilson's *Three Ladies of London* was printed in 1584 'as it hath beene publiquely played', but *Tamburlaine* (1590) is the first to name the company (Admiral's). As we approach the form of words so familiar to us on the titles of Shakespearian quartos, we feel we have entered a new age; and while, as this history will endeavour to show, there was much commerce between the old age and the new, we have in fact entered a new age, the age of the professional actor. When (again in 1584) the Queen's players complained to the Council that the City of London was depriving them of their livelihood, the City maintained that professionalism was as new as it was undesirable, and vainly and speciously supported the semi-amateur stage of old.

It hath not ben used nor thought meete heretofore that players have or shold make their lyving on the art of playeng, but men for their lyvings using other honest and lawfull artes, or reteyned in honest services, have by companies learned some enterludes for some encreasce to their profit by other mens pleasures in vacant time of recreation.

They were wrong in saying that professionalism was new. Actors who were not merely minstrels were touring the country by the mid-fifteenth century and sheltering under the names of honourable persons. But now the professionals were in possession, the future of the drama obviously rested with them, and under powerful patronage they had become almost unassailable and almost respectable.

## 2. *Two Leading Themes*

We may distinguish two leading themes which run through many mid-century interludes, concern with the upbringing of the young and with the evils of social corruption. The advocates of any 'New Learning' are likely to pay especial attention to the upbringing of youth, and a few plays, mostly Edwardian and early Elizabethan, confine themselves to this theme. In theme and in tone they are very different from Redford's *Wit and Science*: their concern is with indoctrination, not with education. The harshness of the doctrine and the treatment advocated do not recommend themselves to our easy-going age, but granted a belief that at the Fall the sparks of grace were wholly extinguished and the Law of Nature abrogated, that without justification by faith man is born to sin as to sorrow, that 'The gayest of us al is but wormes meat' (the only memorable line in *Impatient Poverty*, and the sting is in the tail and depends on St. Bernard's *esca vermium*), that heaven is empty and hell is full, it follows that no allowances must be made for the natural man or the natural child. We have seen that if Wever wrote the lyric in *Lusty Juventus* with the refrain 'In youth is pleasure, in youth is pleasure', he wrote it only to repudiate it.[1] Man is naturally prone to evil from his youth, says the Prologue:

Gyve hym no liberty in youth, nor his folly excuse,
Bowe downe his neck, and keepe him in good awe,
Least he be stubburne, no labour refuse,
To trayne him to wisedome, and teach him Gods law:
For youth is frayle and easy to draw,
By Grace to goodnes, by Nature to yll:
That Nature hath ingrafted, is hard to kyll.

In another Edwardian play, *Nice Wanton*, a mother (Xantippe) has three children Barnabas, Ismael, and Dalila, spoils the last two and does not chastise them as she should, with the result that Dalila dies of the pox and Ismael is 'hanged in chaines and waveth his lockes'. The scenes in which the wicked pair play truant from school and throw dice with Iniquity, and the scene in which Ismael is tried for felony and murder by the incorruptible judge Daniel, are unusually realistic. The difficulty is to make the good characters acceptable: Barnabas is stock and repulsive. 'In that god preserved me, small thanke to you' he observes to his mother. The moral is that parents should not cocker or 'tiddle' their children, or neglect to chastise them, and that children must apply their learning and obey their elders: 'It wil be your profit an other day.' Luke vii and viii gave 'the learned clarke' Lewis Wager the theme for his *Mary Magdalene*. This play, too, may be Edwardian in date, and again there is a striving for realism in the scenes in which Mary, spoiled by parents who 'would not suffer the wynde on me to blowe', rejoices in her finery, her youth, and her pleasure. She is counterbalanced by two model children in two plays of the fifteen-sixties based on stories which had been hallowed by the genius of Chaucer: the heroines in R. B.'s *Appius and Virginia* and John Phillip's *Patient Grissell*. Both are in 'the perte and pricking time of youth', 'when sappie youth, his blossoms did displaye', and both are dutiful and obedient to their parents. How rare such model characters were in life the authors knew very well and presented them as moral *exempla*.

A few plays based on or suggested by the parable of the Prodigal Son may more conveniently be mentioned later,[2] but

[1] See above, p. 39. [2] See below, pp. 96 ff.

two more plays that concern themselves with the nurture of children may be considered here: W. Wager's *The Longer thou Livest, the more Fool thou art* and Ulpian Fulwell's *Like will to Like quoth the Devil to the Collier*, both of the fifteen-sixties. Fulwell observes that 'Sith pithie proverbs in our English tung doo abound' he has chosen one that 'may shew good example'. Wager did the same in his play, and again in *Enough is as Good as a Feast*, and in the lost *'Tis Good Sleeping in a Whole Skin.* There was an early Elizabethan vogue for such titles. George Wapull followed it in *The Tide Tarrieth no Man*, and Thomas Lupton in *All for Money*; and there are plays which survive only in title, like *Far Fetched and Dear Bought is Good for Ladies* (entered in the Stationers' Register in 1566) and *Hit the Nail on the Head.*[1] We have heard much in recent years of recurrent imagery in Shakespeare. In these plays there is a dearth of imagery, but proverbs are plentiful. The sixteenth century regarded them, as did Quintilian,[2] as brief allegories, dark sentences, *aliud verbis, aliud sensu.* Proverbs like 'Young it pricketh that will be a thorn' and 'Best to bend while it is a twig' express a writer's message in one allegorical sentence and when repeated serve as a refrain. Wager tells us that his *Enough is as Good as a Feast* is but a rhetorical amplification of the title.

In Fulwell's play four desperate villains, Tom Tosspot and the swaggerer Ralph Roister his boon companion, Cuthbert Cutpurse and Pierce Pickpurse, come either to beggary or to the gallows, and in their last confessions warn the audience to take example from their fate. Ralph's advice is to 'take time while time is' and use well one's youthful years; Tom blames his parents for not bringing him up to virtue, learning, or an honest trade: Cuthbert and Pierce with halters round their necks implore parents not to allow their children too much liberty. Wager's *The Longer thou Livest* stresses the importance of a good education and especially for those who 'by birth are like to have gubernation In publikque weales'. But the chief character, Moros, is the special case of the fool who is mindless and irredeemable, not perhaps a Christian doctrine. Discipline and Piety and Exercitatio find him past grace, and he is abandoned to his fate. Some of his interests seem harmless enough and (to naturalists, folk-lorists, and lovers of song) even laudable. He knows where to find a red-shank's nest,

[1] On this play, see above, p. 50.

[2] *Inst. Or.* VIII. vi. 57, 58.

but Discipline rebukes him. He can sing songs and ballads like 'Brome, Brome on hill' and 'Robin, lend to me thy Bowe' and Edgar's song in *King Lear*, 'Come over the Boorne Besse'; and (like John Aubrey whom posterity respects and loves) he dreads the time when 'such pretie thinges would soone be gon'; but Discipline tells him he ought to be ashamed of himself. When he adds that he learnt these songs and 'twenty mo' at his mother's knee, he is only strengthening the argument that idleness and the indulgence of parents are responsible for the wickedness of children nowadays.

Another leading theme in these plays is the wickedness of man in society and all the evils of social corruption—social rather than political corruption—and for obvious reasons writers are careful to announce that they make no particular reflections and are loyal subjects of the Crown.[1] Of any radical criticism of society they are innocent. Revolutionary texts of which the Bible provided a store may be found in these plays—'If any man would not work, neither should he eat' (2 Thess. iii. 10)—but they are not pressed home as they are in the radical pamphlets which Robert Crowley addressed to 'the great possessioners' in and out of Parliament during the years 1548–50. And while some of the bitterness of their attacks on the abuses of society may be attributed to disappointment that the 'New Learning' had not produced that land flowing with milk and honey which Bale had prophesied,[2] they do not take up the extreme position of Henry Brinkelow in *The Complaint of Roderick Mors* (*c.* 1548) that but for the faith's sake it might have been more profitable for the commonwealth if lands had remained in the possession of the Church: 'for they did not raise rents nor take such cruel fines as our temporall tyrauntes'. On the other hand, these plays are little tainted with the prudential view that virtue and riches walk hand in hand together, though it does rear its ugly head here and there. In *Impatient Poverty*, which has already been mentioned as a play that goes back to Mary's reign and possibly to Henry's, Peace assures Poverty that he will 'come to rychesse, wythin short space' if he will forsake sensuality and be governed by reason; also that for every penny he spends on the poor God will send him double. Even if we take the author's meaning by the better handle and suppose the riches promised to be spiritual, we cannot give Fulwell the same credit in his

[1] See above, p. 36. [2] See above, p. 37.

*Like will to Like*: the gloss of Virtuous Living upon St. Augustine's *aedes . . . coniunctissimas constituerunt Vertutis et Honoris* (*De Civit. Dei*, v) is that 'Where vertue is, fortune must needs growe: But fortune without vertue hath soon the ouerthrowe.'

In exposing evils in contemporary society these dramatists are doing what Langland had done infinitely better two centuries earlier and Jonson was to do much better a generation or so later. They observed that good tillage land was being converted into pasture, that prices and rents were rising, that the usury condemned by the Bible was never more practised, that 'laws catch flies but let the hornets go free', and they did not stop to enquire whether these were economic necessities due to impersonal causes or to make a distinction between public and private morality. Dramatists and preachers and pamphleteers are moralists to a man, and the difference between the approach of a Langland and a Jonson, or a Nashe or a Dekker, between a medieval friar and a Lever or a Perkins is almost imperceptible. The saying 'Every man for himself and God for us all' was abhorrent to them. While the changing structure of society may lead an age to expose one abuse more than another, and while the rise of an aggressive and flourishing middle class presented problems unknown to Langland, yet most social evils are ever old and ever new. Perjury, pilfery, bribery, flattery, robbery, trumpery, harlotry, and also pitchery-patchery[1] are always with us.

Social criticism in these plays may be illustrated from W. Wager's *Enough is as Good as a Feast*, George Wapull's *The Tide Tarrieth no Man*, and Thomas Lupton's *All for Money*, plays which belong to the period 1564–76. In Wager's play Worldly Man, 'stout and frolike' when the play begins, is converted by Contentation, led astray by Covetousness into believing that enough is not enough, becomes 'a very dunghil and sink of sin', on his deathbed begins to make a will 'In the name of . . .' but can get no further, and is carried off to hell on Satan's back. Two of his victims appear on the scene: a labourer who has received no wages for half a year, and a tenant—'an olde man and speak Cotesolde[2] speech'—who has lived in the same house for nearly forty years and is about to be evicted because he cannot pay a rent racked from £5 to £10 a year, and because Worldly Man desires to convert the house

[1] See above, p. 40. [2] Cotswold.

into a buttery for his drink and anyhow thinks it unseemly that a beggar should live under his nose.

Wapull's play brings upon the stage more examples of man's wickedness and man's inhumanity to man: an old man who has been turned out of his tenement by a greedy landlord, a courtier who borrows money so that he may succeed at court and falls a prey to the usurer and his confederates the scrivener and the broker, a merchant who frequents Paul's Cross at sermon time not for the preaching but to clap debtors by the heels, a bribable sergeant of the Counter, and so on.

Lupton's *All for Money* is even more all-embracing in its presentation of the 'manners of men and fashion of the world nowadayes'.[1] This author was to attack the same abuses in his prose-work *Siuqila.*[2] *Too Good to be True* (1581), the second volume of which is remembered (if at all) because it gives a version of the story of an unjust judge used by Shakespeare in *Measure for Measure.* We may take our choice between the title-page's label 'A Moral and Pitieful Comedie' and the Prologue's a 'pleasant Tragedie': the one is as apt as the other. In the first scene Theology, Science, and Art indict all classes of society (except the poor): bishops and priests, men of learning and science, merchants and artisans—all seek money for private gain. For the rest of his play Lupton requires some thirty characters. There is Money who begets Pleasure who begets Sin (the Vice) who begets Damnation—Mischievous Help, Prest-for-Pleasure, and Swift-to-Sin acting (respectively) as midwives. There are Learning-without-Money who is virtuous, Learning-with-Money whose fate lies in the balance until he comes to the assistance of his fellow scholar, Money-without-Learning who is vicious, and Neither-Money-nor-Learning who is old and pious and lives upon charity. But the indictment of society becomes sharp and particular in the scene in which All-for-Money, dressed like a ruler or magistrate, acts as Money's deputy, and accepts bribes from an even unholier crew of suitors than visited the house of Lovewit in *The Alchemist*: Gregory Graceless a thief and a murderer; a young whore who has murdered her bastard; a man who wishes

[1] While the play requires some thirty-two characters, these, it has been reckoned, could be shared 'among half-a-dozen energetic players'.

[2] i.e. 'aliquis'. The work is a dialogue between Siuqila of Ailgna (Anglia) and Omen (Nemo) of Mauqsun (Nusquam).

to exchange an old wife for a new and cannot persuade the bishop to allow him to abandon the 'olde croust' whom he married for money; Nicholas-never-out-of-law, a rich franklin, who covets a poor neighbour's land; Sir Laurence Livingless, a foolish priest and crypto-Papist, deposed by his bishop for lack of learning and other reasons; old Mother Croote enamoured of 'a holsome young man' whom a 'yong drabbe' has enticed away. The only rejected suitor is poor Moneyless-and-friendless whose crime was to steal a few rags and clothes from a hedge. By implication judges, juries, witnesses, bishops' chancellors are all venal or easily deceived. A Jonathan Swift might have made something of this scene, but Lupton's instrument of war is the battering-ram. For the refinement of irony we look in vain in these plays, or almost in vain. When we find a flash, as in Sin's description of hell, it casts only a momentary illumination:

> Oh, it is a goodly house, it is bigger then a grange,
> It passeth fee simple, for the tytle doeth neuer change.

Examples of the belated use of the morality play are found in the plays of Robert Wilson.[1] If Lupton is severe, he is more severe, and being the better dramatist his blows strike home. We know more about him than we have any right to expect. Contemporaries praise his 'extemporal wit' and often link him with Richard Tarlton.[2] For Leicester's company he wrote *c.* 1581 *The Three Ladies of London*, 'a perfecte patterne for all Estates to looke into'. The three Ladies are Love, Conscience, and Lucre, and they are led astray by the four knaves Dissimulation, Fraud, Simony, and Usury who cuts the throat of Hospitality. The clown is Simplicity. No one is spared except the courtier. Among Wilson's targets are innkeepers and ostlers, pimps and thieves, brewers and brokers, rack-renters, unpatriotic merchants who export necessities and import luxuries, corrupt clergy, dishonest lawyers and judges, foreigners who live 'ten houses in one' and pay high rents. The guests at the wedding of Love and Dissimulation, announced

[1] Robert Wilson is heard of in 1572, 1574, and 1581 as a member of Leicester's company. With Tarlton he joined the newly constituted Queen's company *c.* 1583, and in 1585 received payment on behalf of his company for five plays, one of them a 'pastorall of phillyda & Choryn'. Like Tarlton he had a gift for extemporal verse. He wrote also the lost *Catiline's Conspiracy*, praised by Lodge in 1580 as 'the practice of a good scholler'. In his earlier days he was actor as well as dramatist.

[2] See, for example, p. 157 below.

by Cogging, would be at home in the pages of Bunyan: such as Master Wink-at-wrong, Mistress Deep-deceit, and Ferdinando False-weight. If there is one sentence more than another that sums up Wilson's doctrine, it is that 'Lucre rules the roast'. When the three corrupted ladies and the four knaves are brought to justice, the judge bears the significant name of Nemo. In the sequel, *The Three Lords and Three Ladies of London*, written for the Queen's company in 1589, Wilson pays a tribute to his friend and fellow Tarlton (d. 1588). The same ladies and knaves, clown and judge appear, and are joined by three Lords, Policy, Pomp, and Pleasure. The play is as allegorical but with more 'morall observations' and less censure. Patriotic pride is stirred by the defeat of three Spanish lords (with an Irish herald), and civic pride by the discomfiture of three lords of Lincoln. The shields and impresas of the lords, the costumes of the actors, and the wedding procession provide spectacle and assist the allegory. Significant is the abandonment of the fourteeners, twelves, and tumbling verse of the earlier play for rhymed decasyllabics and blank verse with occasional prose for the comic scenes. Significant, too, is the greater use of metaphor. Wilson moved with the times. In his *Cobbler's Prophecy* (1594), probably a court play, allegory is still present but is modified by gods and goddesses and many typical characters. If (as seems probable) he was the man of the same name who was paid by Henslowe between 1598 and 1600 for several plays written in collaboration with Dekker, Chettle, etc., mostly on historical themes, he had abandoned allegory before the turn of the century.

## 3. *The Vice*

Whatever else the Vice may be, he is always the chief comic character. In theory there is no reason why vice should not be put upon the stage with the same seriousness and sobriety as virtue: in practice, however, dramatists, and many a preacher, knew that men and women will not listen for long to unrelieved gravity. In one of the earliest moralities to survive, *The Pride of Life*, promise is made 'of mirth and eke of kare', and again and again in later Tudor times a play is recommended as 'not only godlie, learned and fruitefull, but also well furnished with pleasaunt myrth and pastime, very delectable for those which shall heare or reade the same' (*Mary Magdalene*). The Prologue to *Susanna*, when admitting to an occasional wanton word or light gesture, reminds the audience that

> nought delightes the hart of men on earth,
> So much as matters grave and sad, if they be mixt with myrth.

And so, before the Reformation as well as after, the wanton word, the obscene jest, the oaths which (in Philip Stubbes's words about the England of 1583) had grown to such perfection that no part of Christ's body was left untorn, all these are placed in the mouths of the representatives of evil. The dramatists could have pleaded that they were observing decorum, but the effect was that the stricter sort of preacher could find no distinction between plays divine and plays profane, between miracle and morality plays and secular plays. The late fourteenth-century priest who attacked all miracle plays of Christ and His saints on the ground that they made 'pley and bourde' of the miracles and works of God is indistinguishable from the London Protestant ministers who attack plays and interludes in the fifteen-seventies.[1] The views of one Papist and two Protestants of the late Tudor period may be cited. William Rainolds of Douai (*Refutation of Whitaker*, Paris, 1583) rebukes Jewel and Whitaker for 'such kind of jesting' as 'would better become some merie felow making sport upon a stage, with a furred hood and a woodden dagger'. The Protestants are harsher. In the same year Philip Stubbes (*Anatomy of Abuses*) maintains that they are accursed who say that plays are as good as sermons. True, you may learn lessons from them 'if you will learne to playe the vice, to sweare, teare and blaspheme both Heauen and Earth. . . . Who wil call him a wiseman, that plaieth the part of a foole and a vice? Who can call him a Christian, who plaieth the part of a devil?' And in his *Direction for the Government of the Tongue* (1593) William Perkins observed that the scriptures in token of a loathing thereof sometimes omit the name of a vice and use instead the name of the contrary virtue: 'This being true, then by proportion the visible representation of the vices of men in the world, which is the substance and matter whereof playes and enterludes are made, is much more to be avoided.'

In its technical use the word 'Vice' first occurs in Heywood's

[1] See below, pp. 164 ff.

*Love* and *Weather*, where No-lover-nor-loved and Merry Report are each described as 'the Vice'.[1] There is nothing especially vicious about either character: the one is a mischievous mocking practical joker, the other is also a joker and a punster who in spite of his light behaviour is engaged as Jove's factotum. In the earliest documentary references the Vice is associated with the jester. George Ferrers was Lord of Misrule at court during the Christmas revels of 1551–2 when 'my lordes vice or dyssarde' was allowed three suits, all highly particoloured, and 'one vyces dagger' together with a fool's bauble cost three shillings. The dagger was of wood, and as with Feste's reference to the 'dagger of lath' of 'the old Vice' (IV. ii. 134) part of the joke lay in its ineffectiveness. (When Falstaff observes of Justice Shallow (III. ii. 343) 'And now is this Vice's dagger become a squire!', he has also in mind its attenuated shape.) Heywood's are not the only plays in which the Vice is mischievous rather than evil: other examples are the practical joker who gives his name to *Jack Juggler*,[2] and the Vices in some late Tudor plays in which the allegorical convention has almost expired, as in John Pickering's *Horestes*, acted at court in 1567. The Vice here is Revenge who calls himself the messenger of the gods, but Horestes would have sought revenge without prompting, the revenge is held to be justifiable, and the Vice is present as a mischief-maker and a buffoon. Similarly in three 'romantic' plays of the early fifteen-seventies, *Cambyses*, *Clyomon and Clamydes*, and *Common Conditions*, the three Vices Ambidexter, Shift, and Common Conditions are jesters and mischief-makers. Ambidexter, as his name suggests, is a double dealer, Common Conditions calls himself so because his 'conditions' (behaviour, conduct) are neither good nor evil but observe a mediocrity, and Shift is no more than a comic and cowardly servant. And in *Patient Grissell* the Vice Politic Persuasion is a mad wag on extremely good terms with the audience. There is no evidence that a Vice was dressed as a fool, yet it is not surprising that a Jacobean writer should have confounded the Fool and the Vice and substituted for the latter's dagger a fool's bauble:

It was wont, when an Interlude was to bee acted in a Countrey-Towne, the first question that an Hob-naile Spectator made, before hee would pay his penny to goe in, was, *Whether there bee a*

[1] See above, p. 28. [2] See below, pp. 106 f.

*Divell and a Foole in the play?* And if the Foole get upon the Divels backe, and beate him with his Cox-combe til he roare, the play is compleat. (J. Gee, *The Foot out of the Snare*, 1624.)

So far the evidence is all for Chambers's view that the character of the Vice is derived from that of the domestic fool or jester—with some tincture in the later plays supplied from the mischief-making servants in Plautus and Terence. But in a substantial group of plays written and performed during the third quarter of the century, the Vice, while always comic, is not merely comic. He possesses many of the characteristics of the comic vices in earlier morality plays. He is not the devil but he is the devil's disciple, he is responsible to the devil for the seduction of men and women, he is the leader of minor vices, and he meets with a bad end. Scholars have discussed his ancestry in several books and many articles. If all are right, the Vice has more ancestors than can be counted on the fingers of one hand. He is descended from the domestic fool; from the devils and the vices in earlier moralities, characters like Tutivillus[1] and his underlings in *Mankind*; from characters like Mirth, the king's messenger in *The Pride of Life*; from servant boys like Pikeharness in the Wakefield 'Killing of Abel' and Colle in the Croxton *Play of the Sacrament*; from the comic characters in the folk play—the ancestors of the Morris fool, the fool of the Mummer's play, the clown of the Sword play; from the medieval sermon, not merely from its 'characters' of the seven deadly sins and their representatives in contemporary life but from its jests and satirical bent; from the plotting servants of Terence and Plautus; from the creative zest of the actors speaking more than was set down for them. It seems probable that if the expression had not been first used in Heywood and misapplied to characters that are uncharacteristic, we should not have doubted the derivation from *vitium* and the descent from the comic sins in the fifteenth-century and early sixteenth-century morality plays. That the term was invented by the actors is a plausible conjecture. From their point of view there was more 'fat' in this part than in any other: the Vice is the part for the leading comedian. Some support for

[1] Tutivillus is a minor devil, not a vice. The derivation from Latin *titivillitium* ('a mere trifle') is uncertain. The character also appears in the Wakefield 'Judicium'. In Skelton's 'titivil' (*Garland*, 1523) the sense has already been weakened to 'a mischievous tale-bearer'.

this view is supplied by the fact that the Vice is never so called in the text, only on the title-page or in the list of characters or in a stage direction or (in *King Darius* alone) in the Prologue.

The character is more important than its ancestry. There are some twenty plays in which a character is referred to as the Vice, and the list which follows is limited to ten of these twenty. If every character that resembles the Vice but is not so labelled were to be added, the number of plays would need to be extended to include such characters as Hypocrisy in *Lusty Juventus*, Envy in *Impatient Poverty*, Iniquity in *Nice Wanton*, Injury in the fragmentary *Albion Knight*, or Hypocrisy in *The Conflict of Conscience*. This warning should be added: of all these Vices Avarice in *Respublica* least conforms to pattern. Only in his opening lines does he address the audience, he does not speak nonsense verses or wear a dagger, he is far from being nimble, though with some reluctance he takes part in a song, he is more a caterpillar of the commonwealth than a deadly sin, he makes no attempt to corrupt the soul of Respublica, and at the end he does not ride to and on the devil but is handed over to the law, as might happen to any unscrupulous politician upon a change of government. The list then is as follows:

1. *Respublica* (Avarice). 2. *King Darius* (Iniquity). 3. *Mary Magdalene* (Infidelity). 4. *Susanna* (Ill Report). 5. *Enough is as Good as a Feast* (Covetousness). 6. *The Trial of Treasure* (Natural Inclination). 7. *Like Will to Like* (Newfangle). 8. *Appius and Virginia* (Haphazard). 9. *The Tide Tarrieth no Man* (Courage, i.e. 'contagious and contrarious' Courage). 10. *All for Money* (Sin).

These Vices have in common certain characteristics, some of which they share with comic characters in other plays. They are the devil's disciples, and they are irredeemable. For a minor vice there may be some hope, but for the Vice none, and at the end of the play he goes to the devil (2, 7), or he is hanged (4, 8), or he is shut up (for a time) in prison (1, 6, 9), or more realistically he departs to play his role elsewhere in the world (5, 10).

In some of these plays (4, 5, 7, 10) the devil makes a personal appearance. He is the Vice's 'dad' as the Vice is 'mine own boy', and the Vice is on such terms with him as to call him 'crookt nose' (4) or 'botell nosed knave' and 'snottie nose' (7, 10) and other names that would sully this page. Satan was 'as deformedly dressed as may be' (10)—years earlier he had entered in the N-town cycle 'in the most orryble wyse'—and the appearance of his vizard in these plays was vividly remembered by Harsnet (*Declaration of Egregious Popish Impostures*, 1603) when he wrote of little children 'afrayd of hell mouth in the old plaies painted with great gang teeth,[1] staring eyes, and a foule bottle nose'. The Revels accounts for 1574–5 mention dishes for devils' eyes. But the stage devil is no longer the serious, terrible Lucifer or Belial of the earliest stratum of the craft plays and of *The Castle of Perseverance* and *Wisdom*. The terror has gone out of him as it was going out of the fairies. Now he is a figure of fun. (But he remained terrible in many a pulpit, where the torments of hell were not alleviated by a jest.) In two plays (5, 7) the Vice rides off on the devil's back, and here again Harsnet comes to our assistance:

> It was a prety part in the old Church-playes, when the nimble Vice would skip up nimbly like a Iacke an Apes into the devils necke, and ride the devil a course, and belabour him with his woodden dagger, til he made him roare, wherat the people would laugh to see the devil so vice-haunted. This action, and passion had som semblance, by reason the devil looked like a patible[2] old *Coridon*, with a payre of hornes on his head, and a Cowes tayle at his breech.

One Vice (10) begs a piece of this tail for a fly-flap. The devil and the Vice have this trait in common that they enter roaring or laughing with a 'Ho, ho, ho' or a 'Ha, ha, ha'. It is their signature tune. So the devil in the facetious epitaph attributed by tradition to Shakespeare: 'Oh! oh! quoth the Devil, 'tis my John-a-Combe.'

The number of ways of raising a laugh in low comedy appears to be limited, and the methods of the Vice resemble those of any bottle-nosed comedian of today or yesterday or any day. Whatever his ancestry may have been, his jokes descend from innumerable sources. Anything served as grist for this comic mill. His first duty was to establish intimacy with the audience. In many plays he was given an opening soliloquy (1, 2, 3, 7, 9: in 4 the devil speaks fifty lines before he is joined by the Vice), although soliloquy is hardly the word to use for speeches that are at times one-sided dialogues with the audience. Sometimes he will address a particular section of the audience: gentlewomen, wives, virgins that long for a husband,

[1] *gang teeth*, large projecting (gag) teeth.
[2] *patible*, capable of suffering something.

husbands married to shrews (4, 6, 9). Or he will distinguish an individual—a gag that never fails to embarrass the victim and delight the audience—with such words as 'How say you woman, you that stand in the angle?' or 'good gentle boy, how likest thou this play?' or 'How say you, little Meg?' (7) or 'How now mayster, how fare you now? How do you synce I was laste with you?' (2). In several plays he greets 'cousin cutpurse' and advises the spectators to look to their pockets (7–10, *Cambyses*, *Horestes*). He is apt to stand upon his dignity and to pretend to be offended if the audience or the minor vices do not defer to him. It is usual for a quarrel to spring up and for a scuffle to follow. In this scuffle, no doubt, his wooden dagger, to which there are repeated references, came in useful: Ben Jonson observed that he would snap with it at everybody he met (*The Staple of News*). In one play (9) he is told to fight 'to prolong the time, while Wantonnesse maketh her ready' in the tiring-house for the part of Greediness. Nimbleness of body was necessary to him not merely for scuffling but for dancing (as with the Vices in *Mankind* a century earlier); and he was required to sing solo or share a part-song with his fellow-Vices or his victims (1–3, 7–9) and in one play to play an accompaniment 'if he may' upon the gittern (7). In *Susanna* (4) the actors apologize for their play because 'we cannot bewtify the same with musickes song'. A song is often a convenient way of closing a scene.

Nimbleness of speech was as necessary to him as nimbleness of body. His tumbling verse cannot but be spoken nimbly.[1] He plays upon words and is given to wanton misprision of another's meaning. This aspect of his speech seems to have impressed Shakespeare, for in *Richard III* (III. i. 82) Gloucester observes:

> Thus, like the formal Vice, Iniquity,
> I moralize two meanings in one word;

and in *The Two Gentlemen* Speed's words to the Clown (III. i. 283) may themselves be an example of double meaning: 'Well, your old vice still: mistake the word.' In many a play the Vice speaks nonsense verses (4–9, *Patient Grissell*). This convention was already old. There is a suggestion of it in one of Hickscorner's speeches, and Ignorance in Rastell's *Four*

[1] See below, p. 70.

*Elements* speaks verses beginning 'Robyn hode in barnys dale stode And lent hym tyl a mapyll thystyll'. He may prate of impossible events—'At Black heath field where great Golias was slain' (5) or 'I can remembre synce Noes ship Was made and builded on Salisbury plaine' (6); like the Fool in *King Lear* he may speak a mock prophesy—'When gayne is no gransier, And gaudes naught set by' (8). Or again, he may speak verses that are part sense and part nonsense (4):

> In my best petticote
> Is there a hole,
> My sister burnt it with a cole
> A skeane of silke
> Will not make it whole,
> Will ye by any sand mistresse?

The last line is a street cry, and serves as a reminder of how closely the comic scenes of these moralities keep to the diction and manners of common life:

This was the chief comic tradition still holding the stage when Shakespeare was a young man. The verses which he gives the Clown in *Twelfth Night* show how vividly he remembered 'dad' the long-nailed devil and his son the Vice:

> Who with dagger of lath,
> In his rage and his wrath,
>   Cries, ah, ha! to the devil,
> Like a mad lad,
> Pare thy nails, dad.
>   Adieu, goodman devil.

### 4. *Metre and Rhetoric*

John Rastell, as has been noted,[1] observed that English was as adequate for the dissemination of knowledge as Greek and Latin. A similar statement, which may have been inspired by Rastell, appears in verses prefixed to *Andria*, the first English translation of a Latin play (*Terence in English*, printed *c.* 1530). Here are confidence in the English language as an instrument for the dissemination of knowledge and a resolve to enrich it by borrowings from other languages; but not until the time of Mulcaster (*Elementary*, 1582) do we find a writer claiming

[1] Above, p. 24.

that an English poet has no need to blush for his language but may find in it 'great and sufficient stuf for Art'. The writers of late Tudor moralities, like the non-dramatic poets who were their contemporaries, apologize for the barbarity of their style and seem to assume that true eloquence is denied to them by the very fact that they are writing in English. Wager in *Enough is as Good as a Feast*, admits that his tongue 'hath not so comely a grace . . . as hath the Latin and Greek: We cannot like them our sentences eloquently place', and it only remains to do one's best. The Prologue to *The Trial of Treasure*—which has the greater part of one scene in common with *Enough*—acknowledges that the style is barbarous, 'not fined with eloquence'; and Prologue and Epilogue to *Patient Grissell* proclaim that the Muses have exiled the author from Helicon's spring, that the gods did not garnish him with rhetoric, that the actors' 'doinges' and his metre are rude, and that 'wantyng hawtie skill' he has written 'So simplye as hee coulde'. It appears to be the actors in *Appius and Virginia*, not the author, who explain that this is their first attempt and that they hope to do better next time.

A precise historical treatment of metrical development in these plays is rendered impossible by uncertain chronology, but the general drift is clear. Blank verse may be omitted from this discussion, and also prose. Prose had to wait till the last quarter of the century. Blank verse, as every schoolboy knows, was 'invented' by Surrey (d. 1547), saw the light of print in 1557, and was first used in drama in Sackville and Norton's *Gorboduc* (1561–2).[1] No writer of morality plays used it until Robert Wilson's *The Three Lords and Three Ladies of London* (of 1589);[2] by that time the morality play proper had become moribund and almost defunct as popular entertainment.

Before Skelton, dramatists had already abandoned the elaborate stanza forms of the miracle plays and the early moralities. But Skelton was still experimenting with a great variety of verse.[3] Rhyme royal, a stately measure usually reserved for serious passages, dominant in *Nature*, *Fulgens and Lucrece*, *Calisto and Melibea*, *Lusty Juventus*, and in the belated and undramatic *Conflict of Conscience*, and important in *Magnificence*, tends to disappear or to be relegated to prologue and epilogue. Sixains and octaves are dominant in a few plays, all of them

[1] See below, pp. 133 ff. [2] See above, p. 59. [3] See above, p. 14.

early: sixains in *The Four Elements* and *Godly Queen Hester* and octaves in *The World and the Child* and *Wealth and Health.* More and more as the century grew older, verse settled into quatrains with alternate rhymes and with four or five beats to the line (dominant in *Mary Magdalene*, *The Longer thou Livest*, *The Disobedient Child*, and *The Tide Tarrieth no Man*) but above all and overwhelmingly into couplets.

This is the staple metre of John Heywood's plays. If he seldom found it necessary to employ other metres, the reason is that he was writing for entertainment and his plays are all of a piece. Writers of allegorical plays who found it necessary to distinguish between virtues and vices had a choice of metres at their disposal, but even so some, and the author of *Respublica* among them, were content for the most part with the couplet or quatrain in four or five beats, making the line jauntier for the vices and heavier for the virtues.

This line depends upon the number of beats or stresses, not of syllables. Of the staggering uncertainties which meet any reader or speaker of these lines complaint has already been made.[1] To James VI in his 'Reulis and Cautelis to be observit and eschewit in Scottis Poesie' (*The Essays of a Prentice in the Divine Art of Poesy*, 1584) the verse was known as *tumbling verse*, and he restricted its use in poetry to flyting and invectives. He distinguished it sharply from the 'flowing verse' which had triumphed when he was writing, in which each stressed syllable (which he called 'foot' and marked —) is preceded by one unstressed syllable (which he also called 'foot' and marked ˘). When tumbling verses 'keip ordour', he observed, they have 'twa short, and ane lang throuch all the line', and verses can be found in the English morality plays of which the movement is anapaestic; as in L. Wager's *Mary Magdalene*:

> With the teares of her eies which on them did fall,
> With the haire of hir head she hath wiped the same.

But as James said of Scotch tumbling verse, and might have said of English, most tumbling verses do not keep order. For example, Wager continues:

> Thinking all other clothes therto over vile,
> Horrible in hir sight is hir synne and blame,
> Thinkyng hir self worthy of eternall exile.

[1] See above, p. 15.

What helps to save the verse from complete prosodic anarchy is the rhyme and (with some writers) the caesura ('sectioun' to King James). Sometimes the caesura recurs regularly and is marked by the punctuation. It is easy to scan as four-beat lines this stanza in rhyme royal from the epilogue to *Nice Wanton*: the only difficulty is to keep awake.

> Right gentle audience, by thys Interlude ye may se
> How daungerous it is, for the frailtye of youth,
> Without good governaunce, to lyve at libertye,
> Suche chaunces as these, oft happen of truth
> Many miscary, it is the more ruth,
> By negligence of their elders, & not taking payne,
> In tyme good learnyng & qualities to attayne.

The movement of such verses has been described as cantilever. The caesura is the central girder which supports and unites the two arms.

Accentual verses, with beats varying from two to five and even more, may do very well for comic passages. An actor can rattle them off at any pace he likes, and a skip, a shrug, a wink, or a dig will help him over many a rough passage. But they are no vehicles for tragedy or for high comedy. Nor are we much nearer to these when what King James called 'flowing verses' came into the morality drama. Wyatt and Surrey, observed George Puttenham, 'greatly pollished our rude & homely maner of vulgar Poesie, from that it had bene before' (*Art of English Poesy*, 1589), and their verses and those of other men writing in the new style gained a wider public in Tottel's immensely popular miscellany of 1557. Very rarely before 1557 do we find 'flowing verse' in the drama. If we meet with it for a few lines, as in a speech in *Respublica* II. i.:

> But as the waving seas, doe flowe and ebbe by course,
> So all thinges els doe chaunge to better and to wurse.
> Greate Cyties and their fame in tyme dooe fade and passe
> Nowe is a Champion fielde where noble Troie was,—

we very soon fall again into tumbling verse. But in the early sixties writers of plays with an admixture of morality elements began to borrow from non-dramatic verse the 'iambic pattern, which Wyatt, Surrey, and others had restored to English verse. Poulter's measure (alternate twelves and fourteeners rhyming in couplets), one of Wyatt's most doubtful legacies, we can ignore, for its role in drama was never important;[1] but in *Patient Grissell*, *Cambyses*, *Appius and Virginia*, *Susanna*, *Horestes*, *Common Conditions*, *Clyomon and Clamydes*, the fourteener appears in all its deadly monotony, all metre and no rhythm. We have escaped out of the frying-pan into the fire, out of a wilderness into a prison. As well get a melodic line out of a see-saw or a metronome as out of strict cantilever and strict fourteeners. (They are usually printed as fourteeners; but the regular recurrence of the caesura after the eighth syllable turns them into eights and sixes. Two of these fourteeners taken together make up the 'common metre' of the metrical psalms.) The new verse served for the virtuous characters, for the vicious continued to tumble in accentual verse. Men like Phillip, Preston, and Pickering could have made the Vice's speeches syllabic, but they knew better. Nimbleness of speech was required of the Vice,[2] and no man can be nimble in fourteeners. The springiness of his four- or five-beat lines, varied with an occasional scurry of two- or three-beat lines and a rattle of rhymes, contrasts strangely with the flatfootedness of the new metre. When Jonson revived the Vice in *The Devil is an Ass*, he revived his tumbling verse: 'How nimble he is!' says the admiring little devil to the great devil Satan.

What has been said of the metre of these plays must not be understood to apply to the songs. The lyric and the lighter forms of verse never suffered from the anarchy which overtook more formal verse. Song in particular must keep order if it is to be fitted to a recurring tune. Some of the songs that appear in these plays were clearly written for the occasion. We may be in doubt about the finest of them all, Lusty Juventus's 'In a Herber grene',[3] but there can be no doubt about the four-part song in L. Wager's *Mary Magdalene* with the refrain 'Hoigh (*or* Huffa) mistresse Mary, I pray you be merie'. Whether demonstrably new or not, the songs are always dramatic: the drinking songs in *Like Will to Like* ('Good hostes lay a Crab in the fire, and broil a messe of Sous a',[4] etc.); the part-song with the refrain 'Lady, lady' in *The Trial of Treasure*, which is not Sir Toby Belch's ballad about Susanna, nor the ballad sung by Egisthus and Clytemnestra in *Horestes*, with the same refrain; or the

[1] See below, p. 73, for the only considerable use of it in drama.
[2] See above, p. 65.
[3] See above, p. 39.
[4] *Sous*, [souse], pickle, especially pig's ears.

good nonsense song in *The Longer thou Livest* which 'hangeth together like fethers in the winde', in which Moros sings the first two lines and three Vices join in the last two:

> Litle pretty nightingale,
> Among the braunches greene,
> Geue us of your Christmasse ale,
> In the honour of saint Steven.

But there are 'songs of good life' in Feste's distinction as well as 'love-songs'. The devil's party has all the lively measures, but the virtuous are also given to song. If the curious reader turns to the pious songs in *The Trial of Treasure* and *Like Will to Like* or to Grissell's song (with the refrain 'Singe danderlie Distaffe, and danderlie Ye Virgens all come learne of mee'), he will find the metre as regular, the syntax as clear, the diction as monosyllabic, as in any psalm in Sternhold and Hopkins.

In language and rhetoric as in metre a sharp distinction is to be drawn between the comic passages and the serious. The language of the Vice and the vices is nourished and even manured by slang and by the rich idioms and proverbs of popular speech. No doubt every good proverb, like every good ballad, has behind it an act of original creation, but it does not begin to live until it is upon the lips of the people. And to live upon their lips it must be concrete and vivid and applicable to many a homely occasion. The makers and the users of these idioms and proverbs may never have heard of figures and tropes, yet here are better examples of metaphor and irony than any in the serious passages. The dramatists did not invent them, but found them to their hand. Consider, for example, some of the many phrases in these plays about the gallows: to play *sursum corda*, to promote a man to preach at the gallow-tree, to be a knight of the halter or the highest of all one's kin, to stye[1] in a string or look through a rope or ride the horse with four ears, to make a spring under a perch looking up toward the sky, to hang in chains and wave one's locks, to leap at a daisy—this last the final act of a hardened criminal as he dived off the ladder. In these jests the tongue of the sixteenth-century Englishman was not so far from his heart. London Bridge provided a daily *memento mori*.

In the serious passages there is no great excess of ornate

[1] *stye*, ascend.

diction: as we should expect, more striking examples may be found in late medieval work like *Mankind* ('O Mercy, my suavius solas & synguler recreatory My predilecte spesyall'). Puttenham advised a courtly poet not to use long polysyllables, especially as rhyme-words: 'they smatch more the schoole of common players then of any delicate Poet, Lyricke or Elegiacke' (*Art of English Poesy*, 1589); and anyone who examines the rhyme-words in *Mary Magdalene* will see how addicted Wager was to abstract latinisms like *ornature*, *conglutinate*, *delectatious*, *testification*, especially when he was endeavouring to elevate his style. But of the ludicrously excessive use of 'inkhorn terms' satirized by Thomas Wilson in his *Art of Rhetoric* (1553) there is little. Nor is there much addiction to the 'figures', whether of words or thoughts, and it is remarkable how sparing these writers are of tropes. A late Elizabethan poet thinks in images, but these Edwardian, Marian, and early Elizabethan writers rarely hazard a metaphor. Perhaps we ought to take them at their word and believe that they were not trying to be poetical.

But the writers of the plays which came under the influence of contemporary non-dramatic poetry and of Seneca, whether in the original or in Jasper Heywood's fourteeners (*Troas*, 1559, *Thyestes*, 1560, *Hercules Furens*, 1561), did try to be poetical and with disastrous results. The authors of *Patient Grissell* (before 1566), of *Appius and Virginia* (before 1567), and of *Cambyses* (before 1569 and possibly as early as 1561) are still dull, but they are dull in a new way. Now the stock classical allusions come flooding in, until the reader sickens of the mention of Venus and Cupid, Tantalus and Sisyphus, the 'sisters three' who 'shear the vital thread', and Pluto's 'darksome den'. The clichés and the odd vocabulary of what Professor Lewis has named 'drab' verse also make their appearance. Tract of time, grisly gulfs, sobbing shrieks and sighs, briny streams, trickling tears, tristful sorrow—all are in *Patient Grissell*; clearly, we do not need to go to *Cambyses* for Falstaff's 'King Cambises' vein'. Do we look for hunting of the letter as blatant as Peter Quince's 'He bravely broach'd his boiling bloody breast'? Appius on Virginius does not fall far short: 'Oh curst and cruell cankerd churle, oh carll unnaturall'; or Virginia on Appius: 'Bid him imbrue his bloudy handes, in giltles bloud of mee.' Nor are we far from Pyramus-Bottom when Virginius exclaims: 'O man, O moulde, O mucke, O clay, O Hell, O hellish hounde.' Earlier writers were not so ambitious and went their simple if pedestrian way. That he that never climbed never fell may be true, but it is also true that he that mounts higher than he ought falls lower than he would. Wherever we look in these plays we find examples of the art of sinking in poetry. In passion they are always ludicrous. Sometimes changes in the meaning and association of words make worse what was already bad:

> Yet Father give me blessing thine, and let me once imbrace
> Thy comely corps in foulded armes, & kisse thy ancient face;
> (*Cambyses*)

or (Susanna to her maids):

> Neyther of you both I thinke, have brought me sope and oyle,
> To wash the sweate of from my skin, or rid away the soyle.

But a good poet will triumph over such bad luck. Saintsbury quotes an example from the lament of the Virgin Mary in the Wakefield 'Crucifixion', and it is worth quoting again, for it serves as an example of the good taste and simplicity of much medieval work before the rot set in in the fifteenth century. We have travelled far since this kind of writing was possible, and so far we have fared worse.

> Alas, Dede, thou dwellys to lang! whi art thou hid fro me?
> Who kend the to my childe to gang? All blak thou makys his ble.
> Now witterly thou wyrkys wrang, the more I will wyte thee
> But if thou will my harte stang, that I myght with hym dee
> And byde:
> Sore syghyng is my sang, for thyrlyd is his hyde.[1]

The change in dramatic rhetoric and versification between the fifteen-forties and the fifteen-sixties may be illustrated in brief by a comparison between *Wit and Science* and its rewriting in the early Elizabethan *The Marriage of Wit and Science*.[2] The contrast is seen not in the comic passages, where the *Marriage* preserves the old tumbling verse, but in the serious. In this play and only in this play is there any considerable use of poulter's measure. Even the Vice, Tediousness, speaks in poulter's at his first appearance, though the 'Hoh hoh hoh'

[1] *Dede*, Death; *kend*, taught; *ble*, countenance; *witterly*, assuredly; *wyte*, blame; *But if*, unless; *stang*, stab; *thyrlyd*, pierced.

[2] See above, p. 45.

with which he departs is extra-metrical. The change in rhetoric is as startling. The clichés of the non-dramatic poets of the mid-century (tract of time, heap of haps, griping cares), the abundance of trite gnomic sayings and among them 'In time soft water drops can hollow hardest flint' which so attracted the young Shakespeare, the stiff use of anaphora where five or six successive lines may begin with 'This same is he . . .' or with 'How many . . .' (as in *3 Henry VI*, II. v), the figure of synœciosis by which contraries are reconciled, a figure chiefly associated with the Petrarchan vogue ('My salve and yet my sore, my comfort and my care' or 'I burn and yet I freeze, I flame and cool as fast'), these are some of the characteristics of the new style which appear, though not excessively, in the *Marriage*. The rhetoric which Kyd and the early Shakespeare were to inherit is already to hand.

## 5. *Conclusion*

We are unfortunate in not possessing morality texts that belong to the hey-day of medieval poetry: only *The Castle of Perseverance* and the fragmentary *The Pride of Life* belong to the earliest years of the fifteenth century, and these are among the most impressive. But for two centuries the vitality of the allegorical tradition in English drama remained unabated. Medwall's *Fulgens and Lucrece*, the interludes of Heywood, *Calisto and Melibea*, *Gentleness and Nobility*, these may suggest that the tradition was decaying and was being supplanted by a secular drama, yet unless our records are even more fragmentary and deceptive than we suppose, we have to wait for a quarter of a century before we find such another group of non-academic secular plays.[1] It is as if the Reformation killed the secular play for a time or alternatively gave the morality drama a new lease of life. Even as late as the eighties and early nineties, when the tradition seems at its last gasp, banished from London for the most part and acted by amateurs and by strolling players in nooks and corners of the kingdom, morality elements appear without seeming archaic in *Doctor Faustus* and *Summer's Last Will and Testament*. Robert Wilson was still writing straight morality plays for London productions at least as late as 1589. And in 1602 was printed *The Contention*

[1] Below, pp. 103 ff.

*between Liberality and Prodigality*, probably the play acted at court by the Chapel children at Shrovetide 1601 and possibly a revision of the *Prodigality* offered at court in 1567–8. Fortune is a leading character, and so she is in Thomas Dekker's brand new morality play *Old Fortunatus* acted at court by the Admiral's men on 27 December 1599. If only *A Moral of Cloth Breeches and Velvet Hose* (entered 27 May 1600) had survived, we might have had a late example of allegory harnessed to social comedy with blank verse and prose and all the latest improvements and acted by the Chamberlain's men, the best company in town. An amusing glimpse of a very late polemical morality is afforded us by the State Papers. In 1628 the Papist Christopher Mallory was accused of playing the Devil's part in a play at Sir John York's house in Yorkshire, and in that character carrying off on his back to hell no less a person than King James.

It is customary to speak of the characters in morality drama as abstractions, but this is misleading. For characters particular and individual, indeed, for a Falstaff or a Hamlet, we must not look, and in how few plays outside Shakespeare shall we find characters so particular and so individual! Yet in the very early *The Pride of Life* we have a king and a queen and a messenger, and we have Death. Death, who may be an abstraction to us, was not so to our ancestors. They knew his lineaments well, and even if they did not carry a death's head about with them, they could not go far without seeing Death the skeleton on a painted cloth or a wall. The Devil, too, was no abstraction. He was still walking abroad, seeking whom he might devour. He was never out of his diocese, as Latimer observed, never non-resident. Nor were the seven deadly sins less vivid to their minds. When the morality was dying and almost dead, Spenser was describing them in his *Faerie Queene* (I. iv) with a wealth of pictorial detail, and Marlowe or his collaborator was introducing them into *Doctor Faustus* 'in their owne proper shapes and likenesse'. Men and women who watched the morality drama had the added advantage of seeing the characters dressed in appropriate costumes and interpreted by the living voice. We have seen how in *Respublica* the allegorical characters are given descriptive labels which are not allegorical,[1] and if we turn to *New Custom* we shall find that on the title-page Perverse Doctrine is described

[1] See above, p. 41.

as 'an olde Popishe priest', Ignorance as 'an other, but elder', Cruelty and Avarice as rufflers, New Custom and Light of the Gospel as ministers, Edification and God's felicity as sages. And so they would appear to their audience.

An eye-witness has left an attractive account of a performance of a morality play. The town was Gloucester, perhaps about the year 1570, and the play *The Cradle of Security*.[1] The spectator was Ralph Willis, and he wrote about it in his *Mount Tabor* (1639), when he was 75 years of age. His account is as follows:

In the city of *Gloucester* the manner is (as I think it is in other like corporations) that when players of enterludes come to towne, they first attend the Mayor to enforme him what noble-mans servants they are, and so to get licence for their publike playing; and if the Mayor like the actors, or would shew respect to their lord and master, he appoints them to play their first playe before himselfe and the Aldermen and common Counsell of the city; and that is called the Mayor's play, where every one that will comes in without money, the Mayor giving the players a reward as hee thinks fit to shew respect unto them. At such a play, my father tooke me with him and made mee stand betweene his leggs, as he sate upon one of the benches where wee saw and heard very well. The play was called (the Cradle of Security) wherin was personated a king or some great prince with his courtiers of severall kinds, amongst which three ladies were in speciall grace with him; and they keeping him in delights and pleasures, drew him from his graver counsellors, hearing of sermons, and listning to good counsell, and admonitions, that in the end they got him to lye downe in a cradle upon the stage, where these three ladies joyning in a sweet song rocked him asleepe, that he snorted againe, and in the meane time closely conveyed under the cloaths where withall he was covered, a vizard like a swines snout upon his face, with three wire chains fastned thereunto, the other three end whereof being holden severally by those ladies, who fall to singing againe, and then discovered his face, that the spectators might see how they had transformed him, going on with their singing.

Whilst all this was acting, there came forth of another doore at the farthest end of the stage two old men, the one in blew with a serjeant-at-armes his mace on his shoulder, the other in red with a drawn sword in his hand and leaning with the other hand upon the others shoulder; and so they two went along in a soft pace round about by the skirt of the stage, till at last they came to the cradle, when all the court was in greatest jollity; and then

[1] *Security*, culpable self-confidence.

the foremost old man with his mace stroke a fearful blow upon the cradle, whereat all the courtiers, with the three ladies and the vizard, all vanished; and the desolate prince starting up bare-faced, and finding himselfe thus sent for to judgement, made a lamentable complaint of his miserable case, and so was carried away by wicked spirits.

The Prince in the moral, Willis explains, was the Wicked of the World; the three ladies, Pride, Covetousness, and Luxury (Lasciviousness); the two old men, the End of the World and the Last Judgement. By no effort of historical imagination are we able to see these moralities as a contemporary audience saw them, but we can note that to this young boy the characters were not mere moral abstractions but first and foremost a king and three ladies and two old men. And we may remember, too, that when he had reached the great age (for his time) of 75 he could write: 'This sight tooke such impression in me that, when I came towards mans estate, it was as fresh in my memory as if I had seen it newly acted.'

# III

# TUDOR MASQUES, PAGEANTS, AND ENTERTAINMENTS

The historian of the drama who is not content to be a chronicler is forced to consider one by one dramatic forms which existed side by side and fructified one another. Miracle and morality, comedy and tragedy, sacred drama and secular, all were being acted before the last quarter of the sixteenth century. But before turning from allegorical drama to non-allegorical, whether academic or non-academic, we may examine in this chapter other forms of entertainment most of which contained dramatic elements and left their mark on the form and the staging of drama proper. For no history of the drama would be complete which failed to take account of the mummings, masques, dances, entertainments, and other 'shews' which enlivened the festive seasons, especially Christmas and Shrovetide, of the Tudor courts. Nor may street pageantry be neglected, whether a royal progress through the streets of an English city and especially London, or at its lower level the annual procession of London's lord mayor as he went to take the oath of allegiance at Westminster on Simon and Jude's day (29 October). The drama itself became less and less allegorical as the sixteenth century approached the seventeenth, but allegory never lost its hold upon pageant and masque, though the themes allegorized became more and more secular. In early years ceremonies like tilt and barriers were hardly dramatic, for they lacked the element of speech and dialogue, though many of them are mimetic. The tournament in the early Middle Ages was a school of battle in which actions spoke loudest; but as this developed into the tilt, a joust on horseback with lances, or into barriers, a joust on foot with swords, combat became less bloody, more ceremonious, and more and more spectacular. The age of chivalry may have been dead, but not the trappings. The day of Elizabeth's accession (17 November) was sumptuously celebrated by a tilt, that of 1590 being described in Peele's *Anglorum Feriae*. In time, especially after barriers were moved indoors, these shows became semi-dramatic, as dialogue as well as speeches interpreted the action and commented on the combatants. So it was in Jonson's speeches at Prince Henry's barriers in 1610 when Merlin awoke Chivalry from her long sleep. The drama became increasingly indebted to these and other ceremonies in incident, spectacle, and dance; and no playwright made better use of them than Shakespeare—barriers in *Richard II*, to confine our examples to his early plays, masques in *Love's Labour's Lost* and *Romeo and Juliet* burlesque entertainments in *Love's Labour's Lost* and *A Midsummer Night's Dream*.

Behind some of these court ceremonies lies the drama of the folk, and the folk play is another of the many ingredients in the rich mixture of Elizabethan drama. The ritual behind the drama of the seasonal festivals, promoting the fertility of man and nature, is almost world-wide and goes back, conjecture supposes, far beyond human record. In our period the origins had long since been forgotten, and the practices that survived were celebrated in many a town and village for luck and for fun. Sixteenth-century Protestants, however, suspected their pagan origins, and attacked them as idolatrous, Papistical, and heathen. Philip Stubbes had no doubt that May games and Whitsun ales were survivals of pagan practices, and the Maypole but a 'stinking Ydol', and to him every leap or skip in the morris or any other dance was 'a leap towards hel'. But the people were not so easily deprived of these ancient customs. If we may believe John Aubrey (*Gentilism and Judaism*), writing as an antiquary anxious to preserve the traces of a past age, many folk customs and beliefs were killed by the puritan interregnum. 'Warres doe not only extinguish Religion and Lawes, but Superstition; and no suffimen is a greater fugator of Phantosmes, than gunpowder.' Yet the mummers' play, the Plough Monday play, the Pace-Eggers' play (the latter performed at Easter, *Pasch*), the Yorkshire sword-dance, these or some of them persist to this day, though not without the aid of artificial insemination.

What began with the folk remained with the folk, but for a few centuries the mumming, the sword-dance, the morris were taken up by the court and developed with ever-increasing lavishness and expenditure on scenery, costume, music, and dance until they culminated in the transitory glories of the masque. The detail of this development, learnedly expounded by Brotanek, Chambers, Reyher, Enid Welsford, Glynne Wickham, and others is almost overwhelming, and only a few characteristic examples can obtain brief mention. Full information has survived of the visit of the Commons of London at Candlemas 1377 to the young prince Richard and his mother shortly before his coronation. 'Disguizedly aparailed' they rode to Kennington, 130 of them, impersonating knights, an emperor, a pope, cardinals, and legates ('with black vizardes like deuils appearing nothing amiable'), saluted the company in the hall, threw dice for gold and jewels, the dice being loaded so that the mummers always lost, drank with the company, danced on one side of the hall to the music of minstrels while the spectators danced on the other, and so returned to London. This sort of mumming was still practised at the court of Henry VIII, and folk-lorists have detected a nucleus of folk-custom in the entry of the band of worshippers, with their sacrificial *exuviae*, to bring the house good luck. A simpler sort of mumming or rather disguising, in which the young Henry delighted, is the impromptu visit, where the visitors were masked, dressed maybe as Robin Hood and his outlaws in Kentish Kendal or as Turks or Russians or with blacked faces as 'Moreskoes'. So in *Love's Labour's Lost* the king and his three courtiers, visored, disguised as Russians or Muscovites, and preceded by blackamoors and music, pay the ladies what was intended to be a surprise visit. An historical occasion (1530) immortalized by Shakespeare was the surprise visit to Wolsey's banquet of Henry and his gentlemen, visored and dressed as shepherds—but in garments of fine cloth of gold. Not so dissimilar is the masque in Capulet's house to which uninvited masked guests were admitted who danced with the ladies. Perhaps the practice of the disguisers dancing with the ladies and not merely among themselves was new and slightly shocking in 1512. On Twelfth Night, says Hall, 'the kyng with a.xi. other were disguised, after the maner of Italie, called a maske, a thyng not seen afore in Englande'. Some of the ladies refused to dance 'because it was not a thyng commonly seen. And after thei daunced and commoned together.' In those days, Hall observes, Henry was 'yonge, and willyng not to be idell'.

But these were silent disguisings, silent except for the unrehearsed flirtations between the dancers. We come nearer to the masque and the street pageant in those shows for which Lydgate wrote explanatory speeches to be spoken by a Presenter, perhaps *c.* 1427–30. The characters, as we should expect, are allegorical, with such familiar figures as Prudence and Fortitude, but also Bacchus, Juno, and Ceres offering wine, oil, and wheat in token of gladness, peace, and plenty. A 'disguising' given before Henry VI at Hertford stands out as a blend of debate and farce, Lydgate's Presenter narrating the sufferings of some 'rude upplandishe' husbands, the wives giving their own 'boystous aunswere'. In 1431 elaborate pageants were prepared for Henry on his entry into Paris to be crowned King of France and on his return to London. There is a striking similarity between the London pageants and later royal progresses through the City and even the Lord Mayors' Shows which became habitual from 1535. At each place where the procession halted a pageant impressed upon the people the message of the day with symbolism and allegory, political, moral, and personal expressed in architecture supplemented by the costumes of the performers, by speech and by song, just as did the pageants ordered by City companies from Middleton, Dekker, and Heywood in the sixteen-twenties and thirties.

Court revels did not flourish during the wars of Lancaster and York, but as the chronicler Edward Hall enters upon 'the Triumphant Reign of Henry the VIII' his pages begin to glow with colour as he describes the lavish revels of the young king, revels which equalled or surpassed in splendour the pageantry of the court of Burgundy under Philip the Good and Charles the Bold and vied with the brilliancies of the court of Francis I. The indoor entertainments are now increasingly important, for here we find the rudiments of the masque. The most sumptuous celebrations before Henry acceded to the throne were held in Westminster Hall in 1501 on the occasion of the marriage of Prince Arthur and Catherine of Aragon. One of these is tripartite: a castle on wheels drawn in by four emblematic (and artificial) lions, eight disguised ladies at the windows, and little boys dressed as maidens on the four turrets singing sweetly as the pageant advanced down the hall to halt before the State; a ship that appeared to be sailing on the sea, the shipmen behaving themselves and speaking after the manner of mariners, and emerging from the ship a Spanish princess and two ambassadors (Hope and Desire) with messages of love which the ladies in the castle spurn; a Mount of Love with eight knights who storm the castle. As at the Burgundian and French courts the superficial aspects of chivalry are made much of just at the period when feudalism was dying. Tilts and barriers at which the young Henry excelled were occasions of great display and ceremony as well as manly activity, as indeed they were to be for another century. Castles continue to abound, no longer 'castles of perseverance' but castles of love peopled by Ardent Desire, Scorn, Disdain, etc., and attacked by knights armed with dates and oranges and defended by ladies with rose water and comfits. One more example may be cited, a blend of oration, song, interlude, and masque, in honour of the peace with France in 1527 and the French ambassadors. Here there are many movements: a Latin oration in praise of the peace; a debate by two men—introduced with song by gentlemen of the Chapel—whether riches were better than love; a mock fight in which the issue is debated by armed knights; an old man who judges that both love and riches are necessary to princes; a richly jewelled Mount from which eight lords descend to dance with ladies; a Cave from which eight ladies (including the Princess Mary) issue to dance with the lords of the Mount; six persons dressed as Icelanders who dance with ladies; eight others (the King among them) dressed in Venetian fashion, all visarded and with beards of gold, who dance their fill until the Queen plucks off the King's visor. And then banqueting till break of day. Here is variety without unity, lavish display, but with what taste? In Italy the masque was already fully developed, music, poetry, dancing and the fine arts all made to serve one harmonious whole. A few Englishmen were not ignorant of Italian art whether directly or through the medium of an Italianate France. But if artists like Torrigiani visited us, they did so briefly, and fortunately for us their business was not with the transitory revels. These were prepared by native craftsmen and produced by men like William Cornish. In France Beaujoyeulx's *Circe* or *Ballet Comique de la Reine*, acclaimed as the first French masque in which plot, ballet, music, and song are made to cohere, was not produced till 1581; and England, thanks in part to the Reformation, had to wait longer.

The appetite of the youthful Queen Elizabeth for pleasure was as robust as that of her father. For example, she and the Spanish ambassador attended at Sir Richard Sackville's house on 10 July 1564 a comedy (late in the evening) followed by a masque with gentlemen dressed in the Queen's colours (black and white = *noblesse* and joy), followed by every kind of preserves and candied fruits. She returned to Westminster by water at 2 a.m. of a windy night. Comedy, barriers, or mock tourney, masque and dancing, these might be the courses of one Gargantuan banquet lasting till break of day. The masque, however, undergoes no striking developments, and not until the Gray's Inn 'Masque of Proteus' of 1595 do we find anticipations of all the characteristics of Jacobean masque except the anti-masque. But thanks to the Queen's fondness for summer progresses, the outdoor entertainment flourished exceedingly, entertainments more dramatic than the Tudor masque the main purpose of which was to introduce disguised noblemen and ladies and their dances. (The English climate was not always kind to the many poets who invented these outdoor shows—as at Norwich in 1578 when Tom Churchyard's water nymphs, looking like drowned rats, had hastily to be metamorphosed into land fairies.) As all English masques must yield the palm to *Comus*, so all English entertainments to *Arcades*. But *Arcades* comes at the end of a long line of pastoral shows, shows which spill over, most beneficially, into the drama proper.

The revels at court during the strenuous Christmas and Shrovetide festivities of the Tudors came under the jurisdiction of the Lord of Misrule or Christmas Lord or Prince. He is not to be confused with the Master of the Revels who for the brief period that he existed[1] before 1545—when he became a permanent official—was appointed, like the Christmas Prince, *pro hac vice*. The Lord of Misrule assumed the same kind of licence as the Abbot of Fools in the medieval Feast of Fools, a new year's festival of the inferior clergy who, like the Boy Bishops, for a few days turned hierarchy topsy-turvy, played checkmate with their betters, to make a Christmas saturnalia. The Lord of Misrule about whom we know most is George Ferrers, contributor to *The Mirror for Magistrates*, Lord of Misrule in successive years under Edward VI, and perhaps the last Lord at court, for the office did not survive Edward's reign and the jealousy of the now permanent Master of the Revels. Perhaps also there were moral objections. 'Christ was never anie Christmas Lord', Samuel Bird was saying in 1580, 'he was never Lord of Misrule'. The Inns of Court also had their Lords of Misrule, likewise mock monarchs invested with a brief authority,

[1] [Wickham (*Early English Stages*, I. 277) throws doubt on any such existence.]

with their own court and bevy of fools and the duty of organizing the revels. These Lords have been thought to descend from the ancient practice of electing a mock king who is put to death after his brief reign to propitiate the god whom he represents, just as the extraordinary practice of killing a cat and a fox between the courses of a banquet—as in the Inner Temple Hall in 1561–2—may be a relic of the heathen rite of sacrifice. (What trace of ancient devilry survived in the practice of inducing a natural to 'mumble a sparrow', who can say?) This was thought 'trim sport', as were stocking and pumping: but the Inns of Court could provide entertainment above the level of practical joking. The greatest of all these revels of which the record has survived is the *Gesta Grayorum* of 1594–5 (printed 1688) performed in the reign of the Prince of Purpoole,[1] memorable not only for its 'Masque of Proteus' but, by reason of a lucky hitch in the programme, for the performance of 'a Comedy of Errors', possibly Shakespeare's, by 'a Company of base and common Fellows'. At the universities too, and especially at Oxford, the Lord of Misrule flourished. The only distinction achieved by Jasper Heywood[2] while probationer Fellow of Merton College (1554–8) was as Lord of Misrule. The fullest of all the records of the functions of this official is in *The Christmas Prince* narrating the revels at St. John's College, Oxford, in 1607–8: he had some crowded hours of glorious life with interludes, comedy, tragedy, masques, and many a show and triumph. Anthony Wood blamed the abolition of these Lords on the puritans, but grave benchers and fellows of colleges must share the responsibility.

[1] Purpoole or Porte Pool is Gray's Inn Lane.

[2] Below, p. 130.

# IV

# SACRED DRAMA

## 1. *Academic Drama*

DURING the middle years of the century the drama was beginning to assume the secular cast which nowadays most of us take for granted. Plays on secular themes there had been before,[1] but if by 1540 the drama was still mainly allegorical, by 1580 it had become mainly secular. Perhaps it would be better to say that it had become ostensibly secular, for, allegorical or non-allegorical, it was still didactic. In the one kind the characters are virtues and vices and follies and as such conduct the action: the literal sense is the allegorical sense and there is no other. In the other, the literal sense is the story, and any edification there may be is inherent, for the spectator to take or leave as he pleases or as he is capable. But any dramatist of the time would have insisted on edification, and so might have wished to say of his play what Tyndale said when attacking the allegorization of the Scriptures: 'There is no story nor gest,[2] seem it never so simple or so vile unto the world, but that thou shalt find therein spirit and life and edifying in the literal sense.' Sometimes, indeed, so strong is the emphasis on teaching, many a play that appears to be non-allegorical easily dissolves into allegory. Self-love changes its name but not its nature under the thin disguise of Philautus.

Many traditions assisted in the secularization of the drama: the drama of the folk (in May-games and Whitsun ales and Christmas festivities), the amusements of the aristocracy (in tilt and barriers, in mummings, disguisings, masques), civic pageants and entertainments, the humanistic drama of academic circles (modelled on the plays of Terence, Plautus, and Seneca). The term academic drama may mean a variety of things. It includes the plays written by schoolmasters with a view to improving their pupils' Latin, indoctrinating them with good morality, and (often enough) improving speech and gesture and 'audacity' by private performance before masters

[1] Above, p. 7.

[2] *gest*, narrative of notable deeds.

and parents. It includes the plays written at both universities, one in Greek (John Christopherson's *Ιεφθάε*), most in Latin, and a very few—and these nearly all comedies—in English. It includes the plays acted in their halls by the Inns of Court, that 'third university of the realm'. At the universities and the Inns of Court dramatists and actors were all amateurs in the sense that their labours were their only reward. So too were the dramatists and actors of school plays, though especially in the last half of the century we must make an exception—and it is an important exception—of the choir-boys at the royal chapels of Westminster and Windsor and at St. Paul's, controlled by masters one of whose duties was to train young actors and produce plays at court.[1] When Elizabeth visited Cambridge in 1564 and Oxford in 1566 and 1592, seven of the nine plays presented were in Latin: both she and her successor were able to follow them. But when not on progress in these universities, but in residence at one of the many royal palaces in the neighbourhood of London, the plays acted before her were always in English, even those presented by the Inns of Court. There appears to be no exception—one example (of many) how the native growth of our drama was fostered under the kindly protection of the court.

The greatness of Elizabethan drama derives solely from the plays written by professional dramatists for the popular stages. If all that has survived from the academic drama had perished, the greatness would suffer no diminution, and the drama of Marlowe, Shakespeare, Jonson, and others would still take its place beside the Greek as the greatest the western world has yet produced. The scholars who wrote the neo-Latin plays acted at Oxford and Cambridge either ignore the popular drama or mention it only to despise it. Even William Gager of Christ Church who won fame in the fifteen-eighties as a writer of Latin tragedy, and was a friend of George Peele, while ardently defending the beneficial effects of acting plays in a learned tongue, joined in the condemnation of all professional acting. A historian would be hard put to it to show that the academic drama of this period was in any way influenced by the popular drama: but when we consider how many university men turned professional dramatists in the last quarter of the century, and how many remained ardent theatre-goers, we

[1] Below, pp. 151 ff.

have to allow that something of the tradition with which they became acquainted as undergraduates and participated in as actors or as audience carried over into the popular drama with notable effects on structure, style, and substance.

The subject is vast and the treatment must be summary. Before considering the development of comedy and tragedy, whether academic or popular, in this period, we may isolate one kind of play which had great vogue in Europe and in England, the play which dramatizes some story of the Bible, whether of the Old Testament or the New. These plays are to be distinguished from the miracle and morality, though they often treat the same subjects, in that they are rarely allegorical, rarely popular, follow classical models, and seek after a different structure and a different style; and they differ from the secular plays of the humanists in that they choose, and choose deliberately, divine themes. They are a part of the Christian humanists' revolt, as apparent in poetry as in drama, from the secular bias encouraged by classical influences. Those of most interest to us were written in the fifteen-forties or a little later, before academic tragedy and tragi-comedy became predominantly Senecan.

## 2. *Sacred Drama, Foreign and English*

All Renaissance things begin in Italy, and the earliest humanistic Latin play has been traced back to 1314. Sacred drama came later. An Italian who spent some years in France published in Paris a tragedy on the Last Judgement (*Theocrisis*, 1514). Six years earlier his tragedy on Christ's death (*Theoandrothanatos*) had appeared at Milan. Gian-Francesco Conti (Quintianus Stoa) has the credit of being the first to apply to a Christian subject the style and structure of Senecan tragedy, but it is his only credit. His tragedies have been dismissed as neither dramatic nor Christian. The impulse to write plays on biblical themes, especially for the edification of schools and colleges, came to France, Germany, and the Low Countries a little later. This impulse was both religious and humanistic, and it was as strong with Protestant humanists as with Catholic. Among the most famous of these dramatists were Théodore de Bèze (Beza), a name to conjure with among the Reformers; Thomas Watson, Bishop of Lincoln, who among the vicissitudes of four reigns remained a faithful Catholic; and George Buchanan, *hujus saeculi poetarum facile princeps*, said by Sir James Melville to be 'of good religion for a poet'. Martin Bucer, who did not himself write plays, came to their defence in his *De Regno Christi* (1550), written after he had retired from his stormy ministry at Strassburg to the quieter haven of a Regius Professorship of Divinity at Cambridge. While allowing that comedy could not entirely be confined to scriptural themes, he maintained that tragedy should always be so based. Like Milton he was 'smit with the love of sacred Song' and sought an argument more heroic and more edifying than any the classics could supply. He would not have denied that in craftsmanship, in *orationis elegantia*, much might be learnt from Sophocles, Euripides, Seneca, but he is so whole-hearted in his devotion to the Uranian Muse that he places the whole emphasis on edification.

Many sacred plays were published separately and many in collections, especially at Antwerp, Cologne, and Basel. At Basel were published two collections of *Comoediae et Tragoediae*, one by Nicholas Brylinger in 1540 and one by John Oporinus (friend of Bale and John Foxe) in 1547. All are in Latin. In one or other of these volumes are the most famous of all 'Prodigal Son' plays, *Acolastus*;[1] the *Susanna* and the *Sapientia Salomonis*[2] of Sixt Birck (Betulius), the German schoolmaster of Augsburg; the *Hamanus* and the already mentioned *Pammachius*[3] of Thomas Kirchmeyer (Naogeorgus); and many another. *Terentius Christianus* (1592), a collection of plays by Cornelius Schonaeus, a Haarlem schoolmaster, is of less interest because it is late. At least eight English editions of two of his sacred plays (called 'comedies' because their issue is fortunate), *Tobaeus* and *Juditha*, both *Terentiano stylo conscriptae*, were printed between 1595 and 1691: but they belong more to the history of education than of drama. Passing mention may be made of a play in Latin prose by the Frenchman Galfredus Petrus, *De vita S. Nicolai comoedia*. It was edited by an Englishman Edward Soppeth and printed by Pynson *c.* 1510, one of the earliest examples of modern drama, if not the earliest, printed in England.

Some of these plays by continental humanists stand out by reason of their merit and the fame of their authors. Two writers

[1] Below, pp. 96 and 97 f. [2] Below, p. 92. [3] Above, p. 36.

may be mentioned: George Buchanan and Théodore de Bèze. Early in his career George Buchanan (1506–82) taught at the College of Guyenne in Bordeaux (1541–4), a college with a dramatic tradition of which it may well be proud, for among its dramatists were Buchanan, Muret, and Joseph Scaliger and among its actors Montaigne. For his pupils Buchanan wrote his two tragedies *Jephthes* and *Baptistes*: the one was published at Paris in 1554, the other in London in 1577 by the Huguenot printer Thomas Vautrollier. Buchanan's plays keep the unities, are divided into episodes by choruses that comment on and take part in the action, and exclude violent action by the use of a messenger. *Baptistes* is more open to political interpretation than *Jephthes*, and this was seen by the man who addressed to Charles I in 1642 a verse translation (but with no mention of its being either verse or a translation) entitled *Tyrannical Government Anatomized: Or, A Discourse Concerning Evil Councellors*. *Jephthes* is the more pathetic, and Buchanan, who first translated the *Medea* and *Alcestis* of Euripides into Latin, profited from the resemblance between the fate of Jephthah's daughter and that of Iphigenia. Of very few neo-Latin plays can it be said that the inspiration is Greek rather than Roman. This play and Thomas Watson's *Absolon*, wrote Ascham, were the only ones he had seen 'able to abyde the trew touch of *Aristotles* precepts and *Euripides* examples'. Sidney's tribute to Buchanan's plays was that they 'doe justly bring forth a divine admiration'.

The sacrifice of Jephthah's daughter is one of the greatest subjects for tragedy which the Bible presents. The sacrifice of Isaac is one of the greatest for 'comedy': a popular theme with the writers of miracle plays, it was chosen by Beza as the plot for his only play. *Abraham Sacrifiant* was written for performance by his pupils in Lausanne, immediately after he had turned Protestant, and it was published at Geneva in 1550. Immensely popular, it was the more so because it was in French. The design was to turn his countrymen from the sonnets of Petrarch to the songs of God: his chorus of shepherds no longer celebrate the joys of earth but sing hymns of divine praise. If learning has gone into the shaping of the play, few, if any, academic plays attempted or achieved a style and treatment so simple and so natural. His play is not *ad sermonis elegantiam accommodatissimus* like those of the imitators of Terence, though it agrees with them in being *ad pietatem excolendam accommodatissimus.* Nor is the play polemical, though his Protestantism has been observed in his insistence on justification by faith. Arthur Golding,[1] who was almost as much the translator-general in his age as Philemon Holland was in his, englished the piece in 1577: it remains moving even in his version.

England can point to no academic dramatists whose fame equals Buchanan's or Beza's. The balance might be redressed if we could say that the comedy of Solomon to which Sir Thomas More added parts and about which he wrote to the grammarian Holt in 1501 was Terentian. It was an interlude written for schoolboys, no doubt the very comedy commended in the recently printed *vulgaria* used at Magdalen College School, Oxford, a comedy in which all the actors, especially he who played Solomon, earned merit. This may be the earliest dramatic criticism in English: if so, it begins in praise.

More and his lost comedies we have to pass by in favour of such minor writers as Bale, Foxe, Udall, Grimald, and Watson. Udall's lost *Ezechias* was polemical; for did not Hezekiah abolish the superstitions in Jewry and reform the state of the church according to the law of God? It may belong to the last years of Hezekiah-Henry VIII. Though in English and by an Oxford man it was chosen for performance before the Queen in 1564 by King's College, Cambridge, not so odd a choice if we remember that Udall had been headmaster of Eton. The Latin *Absalon* of Thomas Watson (the Bishop, not Marlowe's friend), a Cambridge man, also appears to be lost.[2] The author would not suffer it to go abroad, says Ascham, by reason of some small metrical faults such as 'perchance would never be marked, no neither in *Italie* nor France'. It was acted at St. John's, Cambridge, about 1540.

As some compensation for the loss of Udall's play time has preserved Nicholas Grimald's *Christus Redivivus, Comoedia Tragica, sacra & nova* (Cologne, 1543), and *Archipropheta,*

[1] Arthur Golding (1536–1606) was a younger son of an ancient Essex family. Through his half-sister's marriage to the sixteenth Earl of Oxford, he became uncle to the brilliant and unstable seventeenth Earl. A staunch Protestant, he left Cambridge without taking a degree, perhaps during the Marian purge. Cecil, Leicester, Mildmay were among those to whom he dedicated translations from the classical historians and from such French reformers as Calvin, Beza, and De Mornay. His famous translation of the *Metamorphoses* seems hardly in keeping with his other work, but the Ovid which he presents is moralized and allegorized.

[2] [But see s.v. Watson in Bibliography.]

*Tragoedia* (Cologne, 1545).[1] The former belongs to his earliest Oxford days and is dated from Merton College: the latter is dated from Christ Church. While it is divided into five acts and into scenes, *Christus Redivivus* was written before our academic drama submitted to the severity of Senecan influence. A long apologetic epistle in which Grimald reports the views of his tutor John Airy makes this clear. Its doctrines are commonplace and for the most part Horatian and Terentian. On the Continent they were already old-fashioned, yet here for the first time they get an extended airing in a play written for an English stage. For example: the metre of comedy, almost that of Terence, is preserved; chronological arrangement in acts and scenes is followed; as is proper in tragi-comedy the first act (the burial of Christ) is sorrowful, the last (the appearance before the disciples) joyful; for variety's sake comic scenes are dispersed among the sad; if time and place are not strictly observed, neither are they in the *Captivi* of Plautus, which also passes from sorrow to joy. The word *decorum* had not yet passed into English, though it was soon to do so, but in the epistle is the claim that *nihil indecorum* is to be found in the treatment of character, theme, time, or place. Add that Grimald allows himself some twenty characters and a chorus, that there is a strong lyrical element, that the diction depends on Virgil more than the Vulgate. Add also that this was not closet drama, for it was acted at Oxford and Augsburg and in part became absorbed in the German text used for the passion play at Oberammergau until 1740.

*Christus Redivivus* may derive in part from the miracle plays on the Resurrection: the sources of *Archipropheta* are wholly humanistic. It owes nothing to the *Baptistes* of Buchanan, but owes something to Jacob Schöpper's *Joannes Decollatus*. Its structure is less 'regular', however, in that John dies in the fourth act, not the fifth, and it is looser. The plot is revealed in a succession of many short scenes; comic characters, among them

[1] Grimald (?1519/20–?1562) of Christ's College, Cambridge (B.A. 1539–40) became a Fellow of Merton College, Oxford in 1542, but did not stay long: he used insulting language (*verba contumeliosa*) to another Fellow. He moved to Christ Church when it was refounded in 1546 and there read lectures on rhetoric: to the Dean of that college he dedicated *Archipropheta*. Chaplain to Ridley in 1552, he is said to have turned his coat in 1553 and again in 1558. He had, said John Foxe, 'more store of good gifts than of great constancy'. See also C. S. Lewis, *English Literature in the Sixteenth Century*.

Herod's jester, give variety; the passionate love of Herod and Herodias is lusciously displayed (*Da, da roganti basium mihi, cor meum*); and the dancing and the head upon the charger (but not the decollation) are all presented upon the stage. More than one critic has not hesitated to call the tragedy 'romantic', and certainly we must wait for years before we find on the popular stage so full a treatment of lustful and illicit love between the sexes and the evils that follow.

One play is of special interest because we know so much about the circumstances of its performance. *Sapientia Salomonis* was acted in January 1566 by the boys of Westminster School in their College Hall. The new statutes of 1560 had stipulated the annual performance of a Latin play by the scholars and of an English play by the choristers, and from that day to this, with few intermissions, the School has acted a Latin play every year. The Queen had attended a performance of the *Miles Gloriosus* in 1564 and she and Cecilia of Sweden were present in 1566. What they saw was an adaptation of the *Sapientia Salomonis* of Sixt Birck,[1] probably the play acted at Trinity, Cambridge in 1559–60. The chief differences are the conversion of Birck's choruses into dialogues between Wisdom, Justice, and Peace, and the enlargement of the role of the medieval folk-figure, the practical joker Marcolph. A bill of expenses has survived to bring the occasion vividly to our eyes: the backdrop of the city and temple of Jerusalem and its towers; colours for painting the children's faces and gold foil for the prologues' garlands; garments borrowed from the Revels for the Queen of Sheba and two yards of broad say for her head; the names pinned above the 'houses' drawn on paper in gold foil and red and black ink; two boxes of 'dredge'[2] 'to cleare the children'; and a real baby, for attendance upon whom the mother received twelvepence. The copy specially prepared for presentation to the Queen was bound in vellum at a cost of two shillings with the Queen's arms and strings of silk ribbon and is now Add. MS. 20061, our only text.

Of plays in English on sacred history, whether academic or popular, few have survived. Some which contain morality elements—*Godly Queen Hester*, *King Darius*, and L. Wager's *Mary*

[1] Above p. 88.

[2] *dredge*, a kind of sweetmeat for clearing the throat. In 1569 the children 'being horse' were given 'butter beere and dredge'.

*Magdalene*—have already been considered. That there were many such plays acted by the people in towns and villages local records show and will show more abundantly when they have been properly searched. And the Reformation did not put an end to them. We hear of a new *Tobit* with elaborate scenery at Lincoln in 1564, of Thomas Ashton's plays at Shrewsbury from 1561,[1] of a *Destruction of Jerusalem* commissioned by Coventry in 1584 from an Oxford don. We hear too of churchwardens in several counties who hired out 'players' gear' in the fifteen-seventies: at Bungay in Suffolk the players' gowns and coats were made out of old copes. The amount of dramatic activity in sixteenth-century Essex down to 1579 is quite astonishing, and there was much collaboration and visiting between neighbouring towns and villages. But while substantial manuscripts like those of the miracle cycles had a chance of survival, isolated plays had almost none.

If we may judge from the titles of plays acted at Elizabeth's court, sacred drama was not in favour there. The choice of Udall's *Ezechias*, as we have seen, was academic, not courtly. Garter's *Susanna*, which has already received all the attention it deserves, was offered when printed (1569) for all to act. *Jacob and Esau* is another matter. It has been attributed to Udall, and if it has to be assigned to a known dramatist he has the best claim. It would be the handsomest feather in his cap. It may have no right to appear in a treatment of academic drama, for it was not certainly written for children; yet like *Roister Doister* it is a very suitable play for children. The subject-matter might tempt us to link it with other plays on biblical themes: with the Wakefield craft play in which Jacob deceives Esau of Isaac's blessing; with the sixteenth-century interludes in which biblical stories are blended with morality elements; with the Latin academic plays; or with so late an English example as George Peele's *David and Bethsabe*. Yet in effect it stands apart from them all.

John Aubrey stated that the play was acted before Henry VIII, but gave no authority. It was entered for publication in Mary's reign (1557). While Calvinistic in tone, it is in no way polemical, nor does it make direct or oblique attacks on the political or social troubles of the day. It bases itself on *Genesis* xxvii, and *Romans* ix, accepts God's deep and unsearchable

[1] Below, p. 154.

judgements without asking wherefore or why, and develops the human situation with lively invention and vivid detail. If instead of writing 'The Historie of Iacob and Esau' this dramatist had chosen to write a play on the history of an English king, he would not have been passed over by the anthologists for many an inferior playwright. Bale may have been technically a professional dramatist, but essentially he was an amateur, and could never have been more: this dramatist has little to learn about playmaking and stagecraft. The construction and the division into acts and scenes suggest the author's acquaintance with the humanistic religious drama. He knows that a play should begin boldly, and he could not have learnt the lesson from Plautus or Terence. The first scene in which Esau and his sleepily reluctant servant Ragau set out by starlight on a hunting expedition is vigorously comic. He follows this up with a quiet scene, a short conversation between two of Isaac's neighbours, so that when Rebecca and Jacob enter in the third scene, and Isaac and his little boy Mido in the fourth, the situation has been fully revealed and in the most natural manner. As in *Roister Doister* each act is divided into several scenes; in Act IV there are as many as twelve; and why not, so long as the action revolves round a fixed point, the tent of Isaac neighboured by those of Jacob and Esau. Like *Roister Doister* the play has broken completely from the allegorical tradition, but unlike 'our earliest English comedy' none of the characters are stock dramatic types.

In the earliest surviving edition (1568) the play is named a 'History' on title, head-title, and running-title: it is the earliest to be so described in print.[1] It is also called 'A newe mery and wittie Comedie', but it is much more than that. The deception of Isaac, the love of Rebecca for Jacob, the anger and grief of Esau, these are grave matters, and the dramatist shows his humanity in portraying them. Yet there is much comic invention, though it is not allowed to get out of hand. For the most part, it is confined to the fictitious characters, Esau's servant Ragau, and the nurse Deborah who brought the twins into the world, our second comic nurse if we count Udall's Madge Mumblecrust as our first; above all, the small boy Mido who waits upon the blind Isaac and Rebecca's little wench Abra,

[1] Alexander Neville's translation of Seneca's *Oedipus* (1563) was entered, but not printed, as 'the lamentable history of the Prynnce Oedipus'.

two children who are comparatively free from the pert precocity which offends modern taste in most representations of children in our early drama. Like Shakespeare this dramatist interprets the remote by the familiar. Instruction is given that the players are to be dressed as Hebrews, but the stage-directions, which are unusually full and descriptive, tell us that Ragau enters with three greyhounds 'or one as may be gotten', and their names are Lightfoot, Takepart, and Lovell, names which one would not be surprised to find in the kennel-books of today. When busy little Abra sets about cooking Jacob's kids, she goes into an English garden and returns with suckory, liverwort, penny-royal, bugloss, borage, and other herbs grown in all well-appointed English gardens; and the broth she prepares is such as 'God almighty selfe may wet his finger therin'. (John Aubrey who could himself be both homely and vivid takes this expression as an example of 'the low ebb poetry was in in those dark times'!) Some notion of the command this writer has of colloquial English may be gathered from the nurse's 'character' of Abra:

> There is not a pretier gyrle within this mile,
> Than this Abra will be within this litle while.
> As true as any stele: ye may trust her with gold,
> Though it were a bushell, and not a peny tolde.
> As quicke about her worke that must be quickly spead
> As any wenche in twenty mile about her head.
> As fine a peece it is as I knowe but a few,
> Yet perchaunce her husbande of her maye have a shrewe.
> Cat after kinde (saith the proverbe) swete milke wil lap,
> If the mother be a shrew, the daughter can not scape.
> Once the marke she hath, I marvell if she slippe:
> For hir nose is growing aboue hir over lippe.

A historian of the drama whose duty it is to read and (when possible) to see everything, however dull or inept, is tempted to overpraise this lively piece. The language is plain and straightforward, except where time has obscured the meaning of its colloquialisms, and there is no attempt to elevate the style with inkhorn terms and amplifications. Seneca in English translation and the rhetoric which was for so long to muddy the current of English verse were yet to come. Moreover, the play has shape and it has (as the interlude cannot have) dimension: in some 1,600 lines the dramatist has room enough to develop his plot and his characters, both serious and comic. It is hardly too much to say that here is a shape sufficient for any play yet to come. There is only one deficiency, but it is serious. The play does not have, and could hardly have had at this period, the lift and compression of poetry. True, the dramatist does not sink, as Preston sinks in *Cambyses* or Phillip in *Patient Grissell*, but then he does not rise when his theme demands that he should rise. In the serious scenes metaphor and heightening are especially missed, and the doggerel couplet would have killed the work of a lesser man. It is in vain to wish that this gifted playwright might have written his play some thirty or forty years later.

### 3. *Prodigal Son Plays*

One group of plays based upon the parable of the Prodigal Son was remarkably popular and merits a special section to itself. Easily detachable from the sacred narrative, and easily considered as a unit which might be expanded at the dramatist's will, as the history of Jacob or St. John could never be, the parable attracted the attention of many educators of the young. The elaboration of St. Luke's 'he wasted his substance in riotous living' into comic scenes make these plays a suitable bridge to the discussion in the next chapter of early comedy.

The impulse came from the humanists of France, and especially of Germany and the Low Countries who sought to 'Christianize' Terence in many neo-Latin plays written for performance by their pupils: men like Ravisius Textor (Jean Tissier, *c.* 1470–1524) who wrote short *Dialogi* for his pupils at the College of Navarre in the University of Paris, George Macropedius of Utrecht (*Asotus*, written *c.* 1510 and printed 1537, *Rebelles*, 1535, *Petriscus*, 1536), Christopher Stymmelius of Frankfurt-on-Oder (*Studentes*, 1549), and above all Willem de Volder (the fuller) of The Hague (humanistically known as Gnapheus or Fullonius), whose *Acolastus* (1529) was widely acted, went through many editions, and was translated into German, French, and English. The emphasis, as in most of the English derivatives, is upon the importance of parental and scholastic discipline, the evils which accompany the spoiling of children by indulgent fathers and (sometimes) mothers, the damage done to youth by unfaithful and vicious servants, the happiness which attends the virtuous and the misery which awaits the profligate. What of Terence (apart from style and metre) would carry over into this didactic drama are the severity or indulgence shown by father to son, the contrasts between the discreet son and the reckless and between the loyal servant and the disloyal, and 'many morall enstructions amongst the rest of his wanton discourses' (Gascoigne's *Glass of Government*, I. iv). The *meretrix* also makes her appearance in scenes of riot sometimes so developed that we may question how far those for whose benefit these plays were written profited from their instructors.

The extant English remains impress by their bulk if not by their quality. The earliest play, probably printed by William Rastell *c.* 1530–4, survives in a fragment of one leaf to which may be given the title 'The Prodigal Son' or '*Pater, Filius, Uxor*'. The prodigal has married a shrew, is destitute, and turns faggot-seller. Unseen, the father observes the misery of the son and the viciousness of the wife. A *servus*, Robin Runaway, has a speech in Skeltonics. It is not surprising that much the same plot is found in the *Disobedient Child* (printed *c.* 1569) by T. Ingelend, for both plays (especially the latter) depend upon a prose dialogue by Textor, with three characters, Juvenis, Pater, Uxor. In Ingelend's play the disobedient son disobeys his rich and indulgent father, marries on borrowed money a vicious shrew who forces him to sell faggots from street to street, returns to his father in despair, and is sent back to his shrew with the words: 'For I am not he, that wyll the retaync.' In one scene the devil of the morality play delights in his handiwork with 'Ho, ho, ho, what a felowe am I?' Here, as in many a morality, the moral for parents is 'Best to bend while it is a twig' and for children 'Obey your parents, ply your book, and worship God daily.'

Much more influential than Textor's brief dialogues were the plays of the Dutch and German humanists in which the parable was treated in five acts on the Terentian model. Of *Acolastus* we have a translation or rather a prose *ecphrasis* by John Palsgrave (*c.* 1485–1554) published in 1540. To Palsgrave as a recorder and preserver of colloquial English, both in his *L'Esclaircissement de la langue française* (1530) and in this version, our debt is great; he had confidence in his native tongue, and believed that it was come 'to the hygheste perfection that ever hytherto it was'. But he has small claim to appear in a history of the drama. His method was to print Fullonius at the top of his page, and beneath to establish a model for the interpretation of Latin by giving several alternative and idiomatic translations of each phrase, at the same time glossing difficult passages and drawing attention in his margins to his author's phrases and 'figures'. He chose for this exemplar of scholastic construe a play which he conceitedly called 'a very curiouse and artificiall compacted nosegay, gathered out of the moche excellent and odoriferouse swete smellyng gardeynes of the moste pure latyne auctours'. In Fullonius, the father (Pelargus) has a good neighbour and counsellor (Eubulus); the only son Acolastus (or 'stroygood') has a wicked friend (Philautus). Acolastus goes into the world with his portion, falls into the clutches of a parasite (Pamphagus) and his pupil (Pantolabus), is fleeced by them and by the courtesan Lais, drives swine for the husbandman Chremes, repents, and returns to a loving father. Pamphagus acknowledges Terence's Gnatho as the chief of his faculty, and many of the names of the minor characters in the scenes of vice are taken from Terence and Plautus because, as Palsgrave says, Fullonius 'bryngeth in persons of lyke condicions'.

These Prodigal Son plays are considered in this chapter rather than in the second because the characters are non-allegorical;[1] but how easily do some of them dissolve into allegory. Philautus might as well be named Self-Love (he is dropped after the corruption of the son is completed), Eubulus Good Counsel, and Lais Lechery. The same shifting boundaries are to be observed in *Misogonus*. This play in four acts, which may have been written by Anthony Rudd (1549?–1615),[2] was pretty certainly acted between 1568 and 1574 at Trinity College, Cambridge, a college which had recently shown its interest in Prodigal Son plays by the performance of *Acolastus* (1560–1) and *Asotus* (1565–6). In *Misogonus* as in *Acolastus* and in other plays (and as in Lyly's *Euphues*) the chief characters are given Greek names which denote their moral qualities. (The minor characters may bear such names as Dick Duckling, Wasp, Isbell Busbye.) The father (Philogonus) has his good

[1] *Lusty Juventus* and *Nice Wanton*, sometimes treated as Prodigal Son plays, have already been considered.

[2] He rose to be Bishop of St. David's in 1594 and might have gone higher if in a sermon of 1596 he had not offended the Queen by referring to her age and wrinkles.

counsellor and neighbour (Eupelas), and he has also two servants, a good servant, Liturgus (the name of the good servant in *Petriscus*) and an evil, Cacurgus. The latter masquerades as a natural and a jester—he is referred to as Will Somer—but the part he plays is much the same as that of the Vice, and when his villainy is unmasked and he is powerless to do further harm, like the Vice in *Horestes* he cries for a new master. The son Misogonus also has two servants, both evil. To the meretrix (Melissa) is added a rake-helly dicing priest of the old school—'none of this new start up rables'. In Plautus and Terence—see especially the *Heautontimorumenos*—the action sometimes depends upon the discovery of a child (usually a girl) separated from its parents since birth: here the lost child is the elder and virtuous brother (Eugonus), whose arrival upon the scene comforts the father and leads to the repentance of Misogonus. So it is the virtuous brother who is the traveller, not the prodigal.

The play is notorious for the elaboration of the scenes of low life and the extreme viciousness of Misogonus and his boon companions. In addition to drinking, drabbing, and dicing scenes, the dramatist introduces scenes of rustic life—a countryman, his wife (midwife to the mother of Eugonus and Misogonus twenty years ago—the year Piper's Hill was a rye field), and her two gossips: the low humours of peasant life pleased a university audience as a decade or so earlier they had pleased in another Cambridge play, *Gammer Gurton's Needle*.[1] In all these scenes we are given a feast of colloquial English. The metre is mainly tumbling verse with occasional fourteeners, and the illusion that we are listening to spoken English is fortified by the use of contractions like 'Ile', 'youle', 'your' (you are), and many another. We may wonder how a dramatist using the diction of common life could have managed without them, yet before the year 1550 or thereabouts such contractions are rare and after that date they come crowding in. We have only to think of 'Ile guild the Faces of the Groomes withall' to realize the value to Shakespeare's line of so treating the small change of speech. In 1585 King James said that words might be cut short and 'hurland ouer heuch'[2] only in flytings and invectives ('I is neir cair'), never if the subject is love or 'tragedies' ('I sall never cair'). The greatest of his English subjects came to think otherwise.

[1] Below, pp. 109 ff.

[2] *hurland ouer heuch*, (as if) tumbling over crags.

The last of the Prodigal Son plays that deserve notice—after 1580 we have better fish to fry than these academic and didactic pieces—is Gascoigne's *Glass of Government* (1575). Gascoigne[1] is praised by Nashe as a kind of Joshua who did not live to enter the promised land, a man who 'first beate the path to that perfection which our best Poets have aspired to'. As a dramatist he deserves more attention for his *Jocasta* and *Supposes*,[2] because the *Glass* is little more than a didactic tract in five acts in which the author is relentlessly preoccupied with the moral. Perhaps residence in the Low Countries from 1572 familiarized him with their Prodigal Son plays, but so far as we know he does not follow one particular model, and (except in the Prologue and the sententious choruses between the acts) he writes in prose, in the moral scenes a plain, lucid prose with only occasional ornament. That this restraint was deliberate is suggested by the doctrine—in moral poetry 'observe *decorum*, for tryfling allegories or pleasant fygures in serious causes are not most comely' (III. iii). The objection may be made that while there is life in the language of the unregenerate, the speeches of the virtuous are unspeakable upon the stage. Gascoigne gave his answer in his Prologue:

> A Comedie, I meane for to present,
> No *Terence* phrase: his tyme and myne are twaine:
> The verse that pleasde a *Romaine* rashe intent,
> Myght well offend the godly Preachers vayne.

[1] George Gascoigne (1535/40–1577) was born at Cardington, Bedfordshire, of a good family and to a good inheritance. He learned at Cambridge 'such lattyn as I forgatt', amused himself at Gray's Inn, and sat for Bedford in the last parliament of Philip and Mary (1558). He lived the life of a country gentleman, and (1561–3) a courtier, but at court met with no success. In 1561 he married a widow, mother of the poet Nicholas Breton, and became involved in many legal difficulties and financial operations often of a shady nature. He returned for a time to Gray's Inn (1564 or 1565) and prepared for this Inn the *Supposes* and (with Francis Kinwelmershe) *Jocasta*, both presented in 1566. In Bedford Gaol for debt in 1570, two years later he joined the first batch of English volunteers to fight for the Dutch against Spanish supremacy, but returned without fame or fortune the same year. A second visit, made while the *Hundred Sundry Flowers* (1573) was going through the press, was not more successful, and he was back again in the autumn of 1574. In his last years he underwent a moral conversion and published *The Glass of Government* (1575) and several other didactic works in verse and prose. In 1575 he contributed to the Queen's entertainments both at Kenilworth and Woodstock. He at last achieved government employment in 1576: sent to observe affairs in the Low Countries, he behaved with 'humanitie' to the English merchants at Antwerp during the sack of that city.

[2] Below, pp. 115, 136 f.

Deformed shewes were then esteemed muche,
Reformed speeche doth now become us best.
Mens wordes muste weye and tryed be by touche
Of Gods owne worde, wherein the truth doth rest.

The scene is set in Antwerp, the sack of which the author was himself to witness one year later. Two fathers have each of them a wicked elder son (Philosarchus and Philautus) and a virtuous younger son (Philotimus and Philomusus). There are also a good servant Fidus, and a bad servant Ambidexter. A model schoolmaster (Gnomaticus) prepares the four boys for a (very Protestant) University of Douai by teaching them their duty to God and His ministers and to king and country. The wicked are led astray by Lais and her aunt, their servant Echo, and a 'roister', Dick Drum, and they continue in their evil courses at the University. There is no return to a loving father: one son is executed at the Palsgrave's Court for robbery, the other whipped almost to death for fornication at Geneva. The virtuous brethren rise as fast as the wicked fall and achieve distinction in the very places where they witness their brothers' punishments. See how the virtuous are rewarded in this world and the wicked punished: this is the immoral moral of the piece. The phrase 'poetical justice' was not invented till the time of Rymer and Dryden; but the doctrine is strictly observed by Gascoigne; and in this world of wealthy burghers the morality is more than a little prudential.

# V

## COMEDY, *c.* 1540–*c.* 1584

IN Italy, as in England, formal comedy begins with the acting of Terence and Plautus, proceeds to imitation, and leads to original works that apply classical technique to the manners of their own country and the custom of their own language. Whatever the genre of Renaissance literature that is being discussed, we have always to observe that it appeared in Italy a century and more earlier than in England. The earliest Italian comedy in Latin is the *Paulus* (*c.* 1390) of P. P. Vergerio, and like most comedy written in Italy it deals with contemporary life, especially university life, and deals with it satirically. Notorious is the *Chrysis* of Aeneas Sylvius (1444) in which (as occasionally but rarely in England) an ostensible moral purpose is swamped by the realism with which low life is portrayed. The appearance of comedy in the vernacular was not long delayed, and when it came—in such works as the *Suppositi* (acted 1509) and *Negromante* of Ariosto and the *Mandragola* of Machiavelli—it did so with a brilliance which far outshone the earliest efforts of English comedy. A fact by no means irrelevant to the development of English drama is the dependence of much Italian comedy on the *novellieri*.

The story of Latin comedy in Germany is in two respects nearer to our own. There is the use of it, as also of the vernacular, for Reformation polemics. More interesting is the early appearance of plays that adopt the five-act structure in the elaboration of farce and fabliau. Early in the sixteenth century, nearly half a century before our *Gammer Gurton's Needle*, Georgius Macropedius wrote (in Latin) three peasant comedies—*Aluta*, *Andrisca*, and *Bassarus*—which are extremely unclassical in subject, though indebted to Plautus in technique.

Comedy in Renaissance Europe, whether Latin or vernacular, was by no means confined to the satirical mode, but before considering romantic models we may turn to the few surviving examples of English vernacular comedy in the mid-century that show familiarity with the structure of Terence and Plautus.

That the earliest records of the acting of a Latin play should refer to Terence—at King's Hall, Cambridge, in 1510–11 and again in 1516–17—is not surprising, for he was as acceptable to the Renaissance as he had been to the Middle Ages. He was also the earliest of the classical dramatists to be printed in England (Pynson, 1495–7). Some of the reasons for his popularity are well stated by a seventeenth-century schoolmaster, Charles Hoole, who gave him first place among school-authors: his comedies are the very quintessence of familiar Latin, yet apt for the expression of most of our anglicisms; his matter is full of morality, a morality applicable to modern society in which the counterparts to his characters are still to be found; he teaches the true decorum of things and words; and the acting of him in open school with appropriate gestures promotes grace in an orator and dispels that subrustic bashfulness and timorousness which in after life drowns many good parts in men of singular endowments. Of Plautus the earliest record on an English stage is in the Great Chamber at Greenwich in 1519, possibly by Colet's St. Paul's boys under John Rightwise or Ritwise. More memorable is the acting of *Miles Gloriosus* at Cambridge about 1522 with John Leland in the audience and three future councillors-of-state in the cast, William Paget, Thomas Wriothesley, and Stephen Gardiner. In 1545 Gardiner, weighed down by cares of state, reminded Paget of the occasion and contrasted the pleasure they took in the comedy in those carefree days with the tragedy they were now acting in, 'in a worlde where reason prevayleth not, lernyng prevayleth not, convenauntes be not soo regarded, but the lest pretense suffiseth to avoyde thobservation of them'. Could there be a more striking example of the lasting impression acting may leave on an undergraduate's mind?

The earliest English translation of a classical play, Terence's *Andria*, had been published *c.* 1530 and may have been prompted by the Rastell circle,[1] but the earliest original English comedy observing a five-act structure was not acted before *c.* 1553. By a royal warrant dated 13 December 1554 extraordinary powers were given to Nicholas Udall.[2] The warrant

[1] Above, pp. 23 ff.

[2] Born at Southampton *c.* 1505, he was educated at Winchester and Corpus Christi College, Oxford. In 1528, when a Fellow of his college, he was suspected of Lutheranism together with Frith and others. Five years later he and his friend John Leland wrote verses and songs for various pageants shown at the coronation procession of Queen Anne. In 1534 he published his first book, *Flowers for Latin*

instructed the Master and Yeoman of the Revels' Office that 'wher as our welbelovid Nicholas udall haith at sondry seasons convenient heretofore shewid and myndeth herafter to shewe his diligence in setting forthe of dialogwes and Entreludes before us for our Regall disport and recreacion' they were to deliver to him 'soche apparell for his Auctors' as he should think necessary. A historian of the drama is not bound to enquire how it came about that the man who in 1550 (*A Discourse of Peter Martyr*) was writing unbridled abuse of the Papists ('a foule stynking puddle of idolatrie and supersticyon to endelesse damnacyon') was so high in the good graces of Queen Mary in 1553–4. It may be granted that Udall was not a Christian hero like Fisher and More or Frith and Tyndale; it does not follow that he trimmed his sails to every passing breeze. Even if we attribute *Respublica* to him, as there is some ground for doing,[1] we are attributing to him a play that attacks the reign of Edward not for heresy but for bad government; and other plays assigned to him—*Jack Juggler* and *Jacob and Esau*—are hardly more controversial in theology (if we except prologues and epilogues) than *Roister Doister*. Perhaps Udall was willing to keep, and be silent about, his faith under Mary. When John Marbeck, the composer and organist of the Royal Chapel, was sentenced to death for heresy in 1543, Gardiner pardoned him with the words 'What a devil made thee meddle with the Scriptures? Thy

*speaking*, a commentary in English on three plays of Terence; became Headmaster of Eton; and was granted the degree of M.A. (The University's stipulation that he translate no more books out of Latin into English appears to have been withdrawn.) In 1537 his boys performed before Cromwell, the only evidence of his dramatic activity during these early years. On his own admission he got into debt, and led a life of prodigality and vice. Dismissed from Eton in 1541 on charges of theft and unnatural vice, he published his translation of Erasmus's *Apophthegmata* in 1542, settled in London, and became an ardent Reformer. Henry's confounding of Papistry he celebrated in a lost play *Ezechias*. Under the patronage of Queen Katherine Parr he translated Erasmus's *Paraphrase upon Luke* (1545), and supervised the publication of the *Paraphrases* as a whole in 1548/9. For the translation of *John* Princess Mary was in part responsible. He had now retrieved his name (but not his fortune), and became canon of Windsor in 1551. In spite of the militancy he had shown under Edward, he was favoured by Queen Mary and employed in the presentation of masques and plays. He appears to have become an instructor of youth in the Bishop of Winchester's household at Southwark, for in 1555 Gardiner bequeathed forty marks to 'Nicholas Udal my schoolmaster'. In the same year he became Master of the King's Grammar School annexed to Westminster Abbey, a school which became more illustrious after it was refounded in 1560. He died in December 1556.

[1] See above, p. 43.

vocation was another way, wherein thou hast a goodly gift'. And Marbeck continued to compose and play upon the organs. Similarly, the Protestant Roger Ascham was shielded under Mary by Gardiner and appointed Latin Secretary; and in the next reign, although Richard Farrant and Sebastian Westcott were Papists, they continued (in spite of protests) to act as Masters of the Royal Chapel and of St. Paul's respectively until their deaths in 1580 and 1582. A bishop and two queens appear to have been willing to condone heresy in men of good parts provided that they were content to stick to their last.

Of the four plays attributed to Udall *Thersites* would do him least credit. Though perhaps written in the first instance for Christmas festivities, it was certainly acted or prepared for performance in October 1537, for the epilogue prays for the king, the young Prince Edward (born 12 October) and his mother (d. 24 October). The main plot is taken from one of Textor's *Dialogi*, printed at Paris in 1530 and acted at Queens', Cambridge, in 1543. In both, the moral is that Great Barkers are no Biters or The Greatest Boasters are no Doers; but whereas the aim of the French Professor of Rhetoric was to train his students in good Latin and good rhetoric, the Englishman's was to entertain. This is the first play we have met with since *Johan Johan* the chief purpose of which was to amuse. How Homer's braggart persuades Mulciber to make armour for every vulnerable part of his body, how so accoutred he dares the heavens and all the heroes of old, how he ignores the sorrowing pleas of his doting mother, is terrified by a snail (*geminis testudo armata sagittis*), hides behind his mother's back when threatened by a soldier, and at the end abandons club and sword and runs away, this is the simple, farcical theme which Textor supplied. The Englishman finds English equivalents for some of Textor's classical allusions and introduces many a pun, in particular one on a 'sallet' (a helmet and a salad) which was later to do service in *2 Henry VI* (IV. x) and other Elizabethan plays. He adds greatly to the coarseness of his play by introducing Telemachus in search of a remedy for the worms. Whatever power the piece once had to amuse, it has lost long since.

Attempts to prove Udall's authorship from internal evidence are unconvincing, and external evidence is not more compelling. It has been called a children's play and identified with the one Udall's Eton boys acted before Cromwell in 1537, but there is no evidence that it was acted by children. Several allusions to Oxford and its neighbourhood, notably one to Broken Heys (now Gloucester Green), suggest that it formed part of the entertainment provided by some Lord of Misrule at an Oxford college. No one will suppose that the college was Udall's Corpus Christi, where even in our day the acting of plays is an innovation. This distinction the interlude may claim: it provides the earliest datable example (with the possible exception of *Wit and Science*) of an epilogue in which the actors knelt to pray for their sovereign. The practice remained popular till the eighties and left its mark on one epilogue of Shakespeare's (*2 Henry IV*). The incongruity sometimes found between the play and the prayer led Sir John Harington at the close of his notorious *Metamorphosis of Ajax* (1596) to refuse to follow the example of 'my L. (      ) players, who when they have ended a baudie Comedy . . . kneele downe solemnly, and pray all the companie to pray with them for their good Lord and maister.'

For Udall's authorship of *Jack Juggler* a better but not conclusive case can be made out. Whether with a dramatist of Udall's stature authorship can ever be proved by internal evidence is doubtful. Marlowe or Shakespeare or Jonson writing at the top of his form produced work that cannot be mistaken for another's, but no man can say *aut Udallus aut Diabolus. Thersites* goes for its source to neo-Latin, but there is this common bond between *Jack Juggler*, *Respublica*, and *Roister Doister* that they depend on Plautus and Terence. In the morality play, of course, the debt is much slighter, and only betrays itself in the division into acts and scenes and more certainly (for the author might have learnt this from neo-Latin sources) in borrowings like that in Avarice's abuse of Adulation: 'youe *ait aio* youe, yowe *negat nego* yowe', where the words are those of Terence's famous parasite Gnatho (*Eunuchus*, 252). In *Jack Juggler*, as the 'maker' announces, the 'ground' is from 'Plautus first commedie'. The main plot of *Amphitruo* was unsuitable partly because it it could not be adapted to English manners and partly because it would not 'well besime litle boyes handelings'; so putting aside altogether the complications which arose from Jupiter's impersonation of Amphitryon, the dramatist substitutes for Mercury the trickster Jack (or Jake) Juggler and for the slave Sosia the page or lackey Jenkin Careaway, and makes what fun he can in his brief interlude of the bewilderment of Careaway and of his master and mistress caused by Jack's successful assumption of Careaway's identity.

Plautus and Terence made no attempt to conceal their debts to the New Comedy of Greece, and in their plays settings, names, costumes, manners, were all Greek. What surprises us about these earliest remains of Anglo-Latin comedy is the success with which they are adapted to English manners. The author of *Jack Juggler* was already experienced in this kind of adaptation; he tells us in his prologue that he delights to follow Plautus's arguments and 'to draw out the same For to make at seasuns convenient pastime mirth & game'. The characters are English middle class—a London gentleman and his wife (Master Bongrace and Dame Coy), their maid (Alison Trip-and-go), and page. The plot and some of the dialogue are from Plautus, yet are wholly English. The stage represents a street, as in Greek and Roman comedy, with characters entering from either side or from the 'house' of Careaway's master, at the door of which the page knocks (*Hic pulset ostium*) as Sosia knocks in Plautus; but it is a London street and a London house. It is as if this patriotic author was claiming, half a century before Ben Jonson:

> Our *Scene* is *London*, 'cause we would make knowne,
> No countries mirth is better then our owne.

Gravity he eschews: he is writing a play 'both wyttie, very playsent and merye' for the Christmas season and for performance by children. Yet his epilogue (in tumbling rhyme royals) is grave enough. The times are 'queasy'. Innocent people are compelled to believe the moon is made of green cheese or the crow is white; and if they refuse they suffer harm and even lose their lives. Might overcomes right, and right must pretend to be thankful. If this is a covert attack on the doctrine of transubstantiation it is indeed a serious moral to read into a merry play about a juggler's deception.

*Jack Juggler* (printed 1562) cannot be dated exactly, though one or two of the oaths (e.g. 'By our blyssyd lady heaven quene') suggest that it belongs to the reign of Mary. (Yet 'by the mass' is common Elizabethan and is found several times in Shakespeare.) A lucky allusion to *Roister Doister* added to the third edition of Thomas Wilson's *Rule of Reason* (January 1554) at once establishes Udall's authorship and a downward limit. By 1554 Udall's Eton days were far behind him, and his Westminster days were yet to come. Perhaps the play belongs to the later months of 1553 when Udall had become Gardiner's 'school master'. The absence of the title-page in the only known copy (printed *c.* 1566) may have deprived us of some valuable information about the production. The title-page of *Jack Juggler* tells us that the interlude was 'for Chyldren to playe', and so does the prologue; but in *Roister Doister* Udall is silent in prologue and epilogue. Every one assumes that the play was written for children, but it *is* an assumption, though a most likely one.

Whether it is 'the first regular English comedy', the first in English to follow the comic models of Terence and to be divided into acts and scenes as were the texts of this dramatist which Udall used, depends, as we shall see, upon the dating of *Gammer Gurton's Needle*.[1] The scene is a street before the house of the widow Christian Custance, and even when her maids spin, knit, and sew they do so in the street before her house. The two chief characters, the braggart lover Ralph Roister Doister and his parasite Matthew Merrygreek, and the servants of Ralph and of the widow whom Ralph so unsuccessfully woos, these are stock characters in classical comedy. Udall's choice of women characters—a widow and her maidservants—may or may not have been dictated by the exclusion of maidens of rank and good character from the action of Greek and Roman comedy. In many a detail of dramatic technique we are reminded of Terence. For example, the play begins, as so often with him, and as in *Respublica* and *Jack Juggler*, with a long soliloquy disclosing plot and some of the characters; again, it is common for B to enter just as A desires his presence and sometimes to linger before revealing himself so that he may overhear A's soliloquy and acquire some useful information. The play is the work of a man who had Terence almost by heart, a scholar-schoolmaster who had made him a part of the furniture of his mind. It is the work, too, of a man who was abreast of the most recent doctrines on the structure of Terentian comedy, and these, as shown by the learned researches of Professor T. W. Baldwin into the theory of five-act structure from Donatus on, could hardly date before the late forties or early fifties. Udall's first two acts are the protasis or introduction to the intrigue. It is complete by the end of Act II when Custance refuses to read Ralph's letter or accept his presents. The epitasis, where the perturbations begin, the thickening of the plot, opens in Act III when the letter

[1] See below, pp. 109 f.

is read or rather misread; and the 'summa epitasis' is the rout of Ralph and his men at the end of IV as they attempt to storm Custance's house. The catastrophe in V results in her reunion with Gawyn Goodluck, and the end is pacification and tranquillity.

While there is a chief debt to the *Eunuchus* with its braggart lover Thraso and his flattering parasite Gnatho, Udall was so deeply read in Terence that it is not sufficient to say that he borrows from this play or that. At the same time we have to bear in mind that some of the characteristics—the word-play, direct address, the interspersal of song—are as English as they are Roman. We have also to remember that here as in *Jack Juggler* the aim is to acclimatize Roman comedy. Thraso gave a flavour to the character of Roister Doister, but the braggart whose deeds fall far short of his words is at least as old as the Herod of the craft plays. Again, Gnatho contributed to the character of Matthew Merrygreek, but the very active delight with which this character entangles Roister Doister in absurd situations reminds us as much of the mischievous Vice of the morality plays as of the classical parasite. (Nowhere, by the way, is Merrygreek called a 'parasite'. The word was still new in English and perhaps as a social type the parasite is more Roman than English.) As for the women, Constance's old nurse, Madge Mumblecrust, may be our first comic nurse, and she and the maid-servants Tibet Talkapace and Annot Alyface show by word and deed that Englishwomen never shall be slaves.

*Gammer Gurton's Needle* vies with *Roister Doister* for the title 'the first regular English comedy'. A brief statement of the problems may be given, even if the conclusion is inconclusive. The earliest surviving edition was published by Thomas Colwell in 1575, but the play had been entered to him *c.* January 1563 under the title 'Dyccon of Bedlam', the first name in the *dramatis personae*. The words 'Played on Stage, not longe ago in Christes Colledge in Cambridge' on the title of 1575 may have been reprinted from a lost edition of *c.* 1563. The play was 'Made by Mr. S. Mr. of Art', and three claimants have been put forward. One of these is John Still of Christ's, later Bishop of Bath and Wells. Staid in youth and staid in age, he is a most unlikely candidate: when Vice-Chancellor in 1592 he signed a letter to Burghley which asserted that 'Englishe Comedies, for that wee never used any, wee presentlie have none'. A likelier candidate is John Bridges, and no one would have disputed his authorship if his name had begun with S and he had been of Christ's not Pembroke. A distinguished churchman and controversialist, he continued to lighten the severest arguments with homely proverb and fable. We may easily believe that as a young man he could have written this *jeu d'esprit*. Moreover, the first and second of the Marprelate tracts record the report that he wrote it. Lastly, there is Henry Bradley's candidate William Stevenson, who took his B.A. from Christ's in 1549/50 and his M.A. in 1553 and was a Fellow from 1551 to 1554. In each year from 1550/51 to 1553/54 he appears in the College records as a writer or producer of plays. In 1553/54 he is styled 'Mr.', the title of a Master of Arts. The words 'in the king's name' (v. ii. 236) taken together with certain asseverations which suggest the old religion may indicate a date at the end of Edward's reign or the beginning of Mary's, too early for Still and Bridges, but not for Stevenson. For the performance of Stevenson's production of 1553/54 two shillings were spent on the 'waits'. Were they the fiddlers whom Diccon orders to 'pype vpp' between Acts II and III? Accept this date and late 1553 for *Roister Doister* and who shall say which of the two was 'our first regular comedy'? It may be added that 1575, the year of the earliest extant edition, was the year in which Stevenson, then prebendary of Durham, was buried in his cathedral.

*Gammer Gurton's Needle* is the most remarkable college play that has come down to us: though academic in origin its subject-matter is consistently popular. *Roister Doister* has been praised for its Englishness, but there the flavour of Terence is inescapable. In *Gammer Gurton's Needle* all is native except the form: it continues, but with a difference, the farcical treatment of low life practised by Heywood and in many morality plays. The difference is that this writer is putting the old wine of comic rusticity into the new bottles of classical form. There is division into acts and scenes, each act ending a stage in the action; there is the deliberate art of crescendo in the scenes of ludicrous passion; and although (as in the rustic scenes of *Misogonus*) the play is written to amuse a sophisticated audience, the author is so bent on decorum and congruity that his learning never betrays itself in diction or allusion but preserves an absolute control over speech, manners, and character. Moreover, everything here is for entertainment, for the display, not for the correction, of manners and morals. (So it is in *Tom Tyler and his Wife*, acted by boys about the year 1560; but that is a slight interlude which dramatizes in Skeltonic dialogue and song a fabliau-like story of a husband who failed to tame his shrew.) To make five acts about so slight a theme as the loss of a needle is some achievement in what that age called 'amplification', yet the characters are few—Gammer Gurton herself and her servants Hodge and Tib, her enemy Dame Chat, the curate, and the bailey, but above all the Vice-like Diccon of Bedlam who sets all at sixes and sevens. Coarse as the play is, it still gives pleasure on the stage to anyone willing to spend an evening with our rude forefathers.

How far in pre-Elizabethan days vernacular comedy had turned from satirical comedy to romantic comedy the surviving records hardly allow us to say. As samples of what we have lost consider some early plays based on Chaucer. We may neglect the *de patientia Griselidis* (? from Boccaccio or Petrarch or Chaucer) and the *Meliboeus* of Bale's friend Ralph Radcliff, written for his boys at Hitchen,[1] for whatever they were they were not 'romantic'. (Perhaps the play on Griselda did not differ greatly in intention from John Phillip's *Patient Grissel* (*c.* 1566) which stressed 'the good example, of her pacience towardes her Husband: and lykewise, the due obedience of Children, toward their Parentes'. But it must have differed greatly in style and metre.)[2] More serious is the loss of 'The Story of Troilus and Pandar' produced before Henry VIII in 1516 by William Cornish, Master of the Children of the Chapel from 1509 to 1523.[3] Also regrettable is the loss of Grimald's *Troilus*, an Oxford play in English known to us only from Bale. No doubt his play was 'Terentian' and Cornish's was not.

Of the third lost play, Richard Edwards's *Palamon and Arcite*, we know more.[4] It was one of several plays produced by him before the Queen in his old college of Christ Church during the progress of 1566. Acted in two parts on successive evenings with elaborate spectacle, it was enthusiastically received. Here was a play that could have owed little to Terence and Plautus, or to Seneca, but treated themes that we associate above all with the

[1] Above, pp. 38 f. [2] Above, p. 72. [3] Above, pp. 22 f.

[4] Richard Edwards (*c.* 1524–66), poet, playwright, and composer, was scholar and fellow of Corpus Christi College, Oxford (B.A., 1544) and student of Christ Church (M.A., 1547). At Christ Church he must have come in touch with Grimald and his *Archipropheta*. A Gentleman of the Chapel Royal by 1557 he was Master of the Children of that Chapel from 1561 till his death. He was also (from 1564) a member of Lincoln's Inn. Some of his poems—among them the well-known 'In goyng to my naked bedde'—were published in *The Paradise of Dainty Devices* (1576).

golden age of our drama: love and friendship, virginity and married life, not to mention exciting accessories like the cry of Theseus's hounds chasing a fox in the quadrangle while the undergraduates in the windows of the Hall shouted 'nowe, nowe' and the Queen cried 'O excellent . . . those boyes are readie to leap out at windowes to followe the houndes'. The marriage of Emilia after her vows of virginity was received *incredibili spectatorum clamore et plausu*: Elizabeth was not offended.

The one play of Edwards that has survived is *Damon and Pythias* acted before the Queen at Whitehall by Edwards's Children, probably during the Christmas season of 1564–5. In its mixture of 'myrth' and 'care' it resembles *Palamon and Arcite*. 'Tragicomoedia' the Merton register calls it recording a revival of 1568, and in his prologue the dramatist styles it a 'Tragicall Commedie'. This Prologue has the interest of being the first statement of dramatic principles in English. The canons are Horatian and elementary, but the doctrine of congruity in character is clearly stated and the word 'decorum' passes into our language. Decorum it is that roisterers in comedy should speak like roisterers and harlots like harlots, and the doctrine may explain why Edwards, blamed for writing 'toying Playes', turned to this serious theme. No specific source has been found, but the outlines of the story are in a hundred writers, and for particular incidents, speeches, and characters he went to Plutarch and Cicero and doubtless many another. He is careful to explain that Dionysius' court is no other court, and since his play is about tyranny, flattery, and false witness as well as fidelity of friend to friend and counsellor to king the precaution was necessary. Low comedy appears in the proverbial figure of Grim the Collier, and two waggish lackeys and plenty of song give the entertainment expected of choir boys. It is much that a play on so moving a theme should have been written in English by a professional producer. Dignity of theme and treatment was what the drama needed and up to a point Edwards supplied them. His sentiments were none the worse if they were sometimes borrowed from Plutarch and Cicero. They are much the worse if expressed in tumbling rhymed verse of anything up to eight beats and in a diction only ceasing to be pedestrian when it is the diction of low life. But the abysses to which the dramatists of this period can sink have already been discussed.[1]

[1] Above, pp. 71 ff.

We have seen that by the beginning of the sixteenth century, comedy in Italy, while still borrowing much from Roman comedy, had become vernacular in language and native in manners. If for half a century and more Englishmen gave little sign of being aware of these Italian models, that is not surprising. England did not become Italianate until Elizabeth's reign was well under way. A comparison between the reception of Ariosto's epic (1516) and Tasso's (1576 and 1581) makes this clear. No mention has been found of Ariosto before the late fifteen-sixties, or borrowing from *Orlando Furioso* apart from a scrap or two in Surrey, whereas the *Gierusalemme Liberata* is no sooner published than its effects upon our poets at once become apparent. Not until the fifteen-sixties with the translation of the *Cortegiano* and many *novelle* and Ascham's attack on the Italianate Englishman does the vernacular literature of Italy come crowding in. By that decade we had become conscious that culturally we were in comparison a backward nation. Richard Argall in an address to the Reader before Gerard Legh's *Accidence of Armoury* (1562) stoutly maintained that in chivalry and heraldry we had nothing to learn, since our traditions went back to Brute. But, he goes on to say,—and it is here that the earliest known use of 'Tramontani' appears in an English book:

> The Italians, (even at this day a people in whom as yet lyc raked the olde sparkes of the Romane glory) call us on this side the Alpes, *Tramontani*. Noting thereby in us, the lacke of civilitie, and of their countrey curtesy, thinking that nurture hath not yet crept over those wast huge hilles. Thus see we by little and little, howe knowledges crept to places erst unknowen: Yet for we are (as pretely noteth the Poet) severed from the worlde, it is thought, the common knowledges came later to us, then to other our neighbours: for our farther distance from the places where artes first sprang.

Until G. C. Moore Smith's discovery at Queens' College, Cambridge, of a paper dated 1546–7 and headed 'New made Garmentes at the Comædia of Lælia Modenas', it was supposed that English dramatists were not ready for Italian comedy of intrigue until the fifteen-sixties and seventies. But here is evidence in the forties of a version, probably in Latin, of one of the most popular of *cinquecento* comedies. The prose comedy *Gl'Ingannati*, the scene of which is laid in Modena and the chief heroine of which is named Lelia, was acted at Siena in 1531 before the academy of the Intronati and was put into French prose by Charles Estienne (*Le Sacrifice*, later *Les Abusez*) as early as 1543. It is in five acts. The play is grounded, like *Twelfth Night*, upon cross-wooings and mistaken identity, in this case the confusion of Lelia disguised in male attire with her lost brother Fabrizio. A version in Latin verse, *Laelia*, the manuscript of which survives, certainly based on Estienne, was acted at Queens' in 1595, but whether dependent on, or independent of, the earlier version does not appear. The performance of 1595 is too late to have much interest for the historian of English drama, but the appearance on an English stage in 1546–7 of a comedy which may be presumed to have preserved the act-divisions of its source and contained so many of the ingredients of Elizabethan comedy not only in its two pairs of lovers but in its comic nurse, its pedant, its many comic servants, is astonishingly early.

For the next example we have to go either to the shadowy John Jeffere or to George Gascoigne—who attempted so many things for the first time and some of them well. Jeffere's *The Bugbears* (*c.* 1565), based mainly on Grazzini's *La Spiritata* (1561), was acted by boys; Gascoigne's *Supposes* from Ariosto's justly famous *I Suppositi* was produced at Gray's Inn in 1566, the same year as his tragedy, *Jocasta*. That he should have chosen so good a comedy is to his credit. We cannot so praise Abraham Fraunce for his Latin *Victoria* acted at St. John's, Cambridge, in 1579, or Anthony Munday (if he is the translator and not merely the introducer) for the English *Fedele and Fortunio: Two Italian Gentlemen* acted at court before 1585, both versions of Luigi Pasqualigo's inferior comedy of intrigue, *Il Fedele*. Also to Gascoigne's credit is that knowing both the prose and the verse versions he preferred to use prose, so presenting us with the first prose comedy in English. Jeffere, while earning merit for much lively colloquial English, rejected the prose of his original for tumbling verse. Gascoigne's prose is vigorous, as his verse could hardly have been, and in spite of a few euphuistic flourishes it is also plain. An example of his speakable English is the translation of Ariosto's ironical comment apropos of a long-lost son restored to his father: 'the strangest case that ever you heard: a man might make a Comedie of it.' The run of the words is not inferior to Fabian's 'If this were played upon a stage now, I could condemn it as an improbable fiction' (*Twelfth Night*, III. iv).

*Supposes* and *Gorboduc* are the Inns of Court's most importan contributions to comedy and to tragedy. The *Suppositi* (the original (prose) version was first acted in 1509) is memorable not because it is indebted to Plautus and Terence, but because being dramatic imitation not dramatic larceny it is an original and well shaped comedy of intrigue, clear as crystal in its design and the work of a great craftsman and a great ironist. It is a play of 'supposes' or mistaken suppositions, a comedy of errors. A young gentleman of Sicily sent by his father to study at Ferrara exchanges identity with his servant that he may enjoy in secret the favours of his mistress, daughter of a gentleman of Ferrara. The action is complicated by an elderly doctor, suitor to the daughter, who turns out to be the father of the servant, and it is clarified by an inquisitive parasite whose appetite for acquiring and disclosing family secrets conveniently rescues the solution from too much embarrassment. At the play's end the two fathers bless the marriage of the young lovers. It has often been stated that the manners of Roman comedy were more agreeable to Italian manners than to English. Strange to English comedy and intolerable to the manners (or stage manners) of that time is the fact that in Gascoigne and Jeffere the heroine is the mistress of the hero. Strange and as intolerable is the fact that in Gascoigne the lovers do not meet upon the stage before the closing lines and never speak to each other. But in other respects there is much in *Supposes* that might be a model for the author of *The Comedy of Errors* and *The Taming of the Shrew*. The one important addition Gascoigne makes is to the speech in which the father laments the dishonour of the daughter. The grief is there in Ariosto as it is in the *Captivi*, as it is in *The Comedy of Errors*; and it adds a deeper note. But Gascoigne adds a moral, the moral of godly education and obedience to parents.[1]

But what of romantic comedy? Where are the anticipations before Lyly and Greene and the young Shakespeare of the comedy based upon the love between the sexes, the settled mode of the great age of Elizabethan comedy? It is extraordinary how few traces have survived. We shall not look for this sort in the followers of Plautus and Terence or in the translators or adapters of Italian comedy. Italian works, however, were being imported in the fifteen-sixties and seventies which would draw any

[1] Above, pp. 52 ff, 96 ff.

susceptible mind to a consideration of courtship and marriage, and all the refinements of love. The roof and crown of all such books is Castiglione's *Il Cortegiano* (1516) which waited till 1561 before it was put into a rough but serviceable English dress by Thomas Hoby; and other works soon followed like George Pettie's *Civil Conversation* (1581) from Stefano Guazzo and Geoffrey Fenton's *Monophylo* (1572) from Étienne Pasquier. But the first writer fully to profit from these books of courtesy was John Lyly.

When our dramatists began to borrow from the Italian *novelle* is uncertain. There is some evidence that they did not wait until the collections from Boccaccio, Bandello, and others, by William Painter and Geoffrey Fenton, were published in 1566 and 1567, 'fonde bookes' as Ascham called them, 'sold in every shop in London, commended by honest titles the soner to corrupt honest maners', foreign delights, Stephen Gosson maintained in 1582, which poisoned the old manners of our country when read, and when acted swept whole cities into the devil's lap. Perhaps the dramatists did not make much use of them before the fifteen-eighties. The earliest University play based on a *novella* is the Latin *Hymenaeus* acted at St. John's, Cambridge, in 1579 and written perhaps by Henry Hickman who played the hero or Abraham Fraunce, the hero's father. The play is a very free treatment of *Decameron*, IV. x, a story not at that date translated into English. There is much invention, and there is some accommodation to English manners. In both a sleeping potion taken in error by the lover is responsible for the main action, but Boccaccio's rascally lover becomes a young student of Padua, and the adulterous wife is changed to a young girl to whom the lover is happily betrothed at the end. The play has been styled the first English romantic comedy, and while the lovers only converse together in one scene, there is sympathy for their predicament, as also for the sorrowing father who believes his son to be dead. Like Romeo the lover is forbidden his mistress's house and secures entrance in the disguise of a masquer.

George Whetstone's *Promos and Cassandra* (2 parts, 1578) may never have been acted.[1] In a marginal note to the prose version

[1] Born *c.* 1544 of wealthy burgher family, he was probably at one of the Inns of Court in 1576. He associated with George Gascoigne (whose life he wrote in 1577). He sailed with Gilbert in an abortive expedition of 1578, and travelled to Italy in 1580. In 1587 he was stabbed to death by an English captain while serving in the Low Countries.

in his *Heptameron* (1582) he tells us that it had not yet been presented upon the stage. He took the most unusual step of publishing the play himself and providing descriptive directions for the reader, if not the actor. The plot he took from Cinthio's *novella* rather than his play *Epitia.* There is much low comedy, the language of which Whetstone excuses on the plea of decorum, and the verse varies from fourteeners through blank verse to tumbling verse. Here is a plot for romantic tragi-comedy, the plot of *Measure for Measure,* but with Whetstone the moral is conventional and overstressed and the treatment inert. A man might read the play and conclude that nothing could come out of it, yet (as the Elizabethans never tired of saying) a bee sucks honey out of the bitterest flowers; and out of a little spark came a great flame.

In his dedication Whetstone makes the interesting comment that whereas the Italian (and following him the Frenchman and the Spaniard) are lascivious in their comedies, and the German too holy, presenting 'on everye common Stage, what Preachers should pronounce in Pulpets', the Englishman grounds his work on impossibilities, in three hours running through the whole world, marrying, begetting children, making children men, fetching devils from hell, making clowns companions of kings, and all this to make the people laugh. The passage reminds us of Sidney's attack some four years later on English comedy and tragedy, but with Sidney the attack is in part an attack on tragi-comedy and in part an insistence on the unities of time and place. In the kind of comedy which Sidney advocated, and later Jonson, a kind approved by all orthodox Renaissance critics and developed from Greek and Roman practice, there was no room for romantic comedy. Comedy is 'an imitation of the common errors of our life' represented 'in the most ridiculous and scornefull sort that may be'. It handles 'private and domestical matters'. It should not 'match Horn-pypes and Funeralls' but should eschew 'mongrell Tragy-comedie' as a form which achieves neither the sportfulness of comedy nor the 'admiration and commiseration' proper to tragedy. It will prefer delightful teaching, the end of poetry, to matters that serve only to lift up a loud laughter. Sidney was writing *c.* 1582 immediately before the dawn and could find nothing to praise in tragedy but *Gorboduc,* and in comedy nothing at all. But then his point of view was that of a classicist, of a man well seen in the Italian critics of

Aristotle, though well able to stand by his own perceptions and to express his doctrine in a prose that gives constant delight. If he had lived to see and read Shakespeare, it is likely he would have been as generous in praise of his romantic comedy and tragedy as was Jonson.

The work which is thought to have goaded Sidney into writing his critical treatise, Stephen Gosson's *School of Abuse* (1579),[1] dedicated to Sidney without permission and by him scorned, throws much light on the popular stage at a period when the evidence of the plays themselves is almost wholly wanting. Gosson's opinions are the more valuable because he had written plays himself. One of these was 'a cast of Italian devises, called, *The Comedie of Captaine Mario*'; later he was to deplore that 'a greate number of my gay countrimen . . . beare a sharper smacke of Italian devises in their heades, then of English religion in their heartes'. Most surprising is his mention of two 'prose books' acted at the Bel Savage inn or playhouse, 'where you shall finde never a woorde without wit, never a line without pith, never a letter placed in vaine'. It reads like Elizabethan praise of John Lyly's comedies, yet there is no evidence that Lyly turned dramatist before 1584 and then for the children's companies acting at the Blackfriars. Other plays which Gosson found it possible to praise as good and sweet are the *Jew* and *Ptolemy*, both acted at the Bull inn, the former 'representing the greedinesse of worldly chusers, and bloody mindes of Usurers' in a plot which bore some resemblance to that of *The Merchant of Venice*.

In his *Plays Confuted in Five Actions* (1582) Gosson gives a remarkable view of the range of subject-matter in plays acted on the public stages. Allow something for a pamphleteer's inflation of his case, and even so enough is left to suggest that the lost material might wholly change our estimate of the drama of this period:

I may boldely say it, because I have seene it, that the *Palace of pleasure*, the *Golden Asse*, the *Æthiopian historie*, *Amadis of Fraunce*, the

[1] Stephen Gosson (1554–1624), born at Canterbury ten years before Marlowe, was educated at the King's School and Corpus Christi College, Oxford. In 1577 he came to London and turned playwright and (perhaps) actor, probably for Leicester's company. Three of his plays are known by title: *Captain Mario*, *Praise at Parting*, and *Catiline's Conspiracy*. But he repented, attacked the stage and other social evils in *The School of Abuse*, defended the *School* in three more pamphlets against such opponents as Thomas Lodge, tried tutoring in the country, travelled as far as Rome and back again, and in 1584 took holy orders. For twenty-four years he was rector of St. Botolph's, Bishopsgate.

*Rounde table*, *baudie Comedies* in *Latine*, *French*, *Italian*, and *Spanish*, have beene throughly ransackt, to furnish the Playe houses in London.

For any check on Gosson's statement or any specific knowledge of the comedies which he and Sidney were attacking we should indeed be badly off were it not for the mercy that for a few years between 1567 and 1585 the yearly accounts of the Revels Office preserve not only elaborate details of the expenditure on scenery, costumes, and stages, but the titles of the plays acted at court. A decade before Lyly turned playwright the children's companies were acting plays based on classical legend: among these were *Ajax and Ulysses* (1572), *Narcissus* (1572) with the exciting spectacle of a fox let loose in the court and chased by hunters with 'howndes, hornes, and hallowing' and some expensive thunder and lightning, *Alcmaeon* (1573), *Perseus and Andromeda* (1574), *Cupid and Psyche* (*c.* 1582). Other children's plays are on some theme from classical history; these it will be convenient to consider later.[1] The plot of one play was based on medieval romance, *Paris and Vienne* acted by the boys of Westminster on Shrove Tuesday at night, 1572. Based on Caxton's edition of that fair medieval legend, it could not have failed to show Vienne and her faithful confidant disguising themselves in man's clothing and both lovers counting the world well lost till their constancy was rewarded. In tourney and barriers Paris won the crystal shield for Vienne, the children being supplied with hobby horses. In passing it may be said that a choir-boys' play which has survived in manuscript, *Juli and Julian*, of unknown origin and uncertain date (?*c.* 1570), can hardly be called romantic comedy. The crafty servant who helps his young master (Juli) to marry his mother's maid (Julian) descends from Roman comedy, and the manners are sometimes more Roman than English. Another play, *Theagines and Chariclea* (1572–3), whether acted by children or adults we do not know, was based on the romance of Heliodorus, but on which part we are left to surmise from the Revels Office order for two spears and an altar for Theagines. The altar was no doubt that sacrificial altar of heated golden bars in the tenth book (Underdowne's translation was published in 1569) which through the chastity of the hero and heroine proved powerless to injure them. Another source of romantic fable is suggested by the title

[1] Below, p. 146.

*Titus and Gisippus*, acted by the children of Paul's in 1579: it is the story of romantic friendship told in the *Decameron*, x. viii and also by Sir Thomas Elyot in *The Governour* (1531). More interesting when we remember *Much Ado About Nothing* is the *Ariodante and Genevra* presented by Mulcaster's boys in 1583. The story is of the deceit by which Ariodante (Claudio) is deluded and of how the good name of Genevra (Hero) is cleared. Peter Beverley's rendering into English verse, the first of any part of Ariosto (*Orlando Furioso*, v) had appeared in 1566.

More exciting is the list of plays presented at court at this time by the adult companies performing under the patronage of the Earls of Leicester, Suffolk, Warwick, and other noblemen. By the middle of the fifteen-seventies the adult companies, if we may judge from the number of their performances at court, were for the first time outstripping the children, and before the end of the decade the first two permanent theatres in England, the Theatre and the Curtain, had become available to them.[1] It is not too much to say that for the historian of the English drama the centre of interest shifts for the first time from the universities, the Inns of Court, and the schools to the professional companies of adult players. Even a partial list of the comedies which they acted at court in the fifteen-seventies is impressive. It will be seen that if the children on the whole preferred plays from classical history and legend, the adult players preferred romantic drama:

*Cloridon and Radiamanta* (1572). *Mamillia* (1573). *Preda and Lucia* (1573). *Herpetulus the Blue Knight and Perobia* (1574). *Panecia* (1574). *Philemon and Felicia* (1574). *Pretestus* (1574). *History of Cynocephali* (1577). *Irish Knight* (1577). *History of the Solitary Knight* (1577). *Three Sisters of Mantua* (1579). *Duke of Milan and the Marquis of Mantua* (1579). *Knight in the Burning Rock* (1579). *Rape of the Second Helen* (1579). *Soldan and the Duke of* —— (1580).

Not all of these plays may be romantic, but many are, and some may be plausibly guessed to be derived from medieval romances of chivalry and adventure. The translations into English prose of the Amadis de Gaule cycle and the like date mainly from the fifteen-eighties and nineties, but attacks on the reading of medieval romance are frequent before 1580. Vives would have prohibited by law the perusal of such works as *Paris*

[1] Below, pp. 158 ff.

*and Vienne*, and *Amadis*; he and others attack the Arthurian romances and the immensely popular and seemingly innocent *Bevis of Hampton*, *Guy of Warwick*, as well as such essential elements of English folklore as Friar Rush, Adam Bell, and Robin Hood. The people who were the support of the professional theatre thought otherwise, and so, it would seem, did Elizabeth and her court. Among the plays in the above list which may be assumed to depend on romances of chivalry are *Herpetulus the Blue Knight*, *The Irish Knight*, *The Solitary Knight*, *The Knight in the Burning Rock*. The Irish Knight, it has been suggested, may be Morhoult of Ireland who belongs to the Arthurian cycle; *Herpetulus* contained a Vice, Diligence, and a dragon; *The Rape of the Second Helen* probably, and *The Solitary Knight* possibly, depend upon Feliciano de Silva's *Florisel de Niquea* which had been translated into French in 1553; and *The Knight in the Burning Rock* was taken from the late Spanish romance *Espejo de Principes y Cavalleros*, Margaret Tyler's translation of which was licensed in 1578 and published as *The Mirrour of Princely Deeds and Knighthood*. Other plays in the list like *Philemon and Felicia*, *The Three Sisters of Mantua*, and *The Duke of Milan* may be of Italian origin, and it has been suggested that Cloridon and Radiamanta are a scribe's perversion of Clodion and Bradamante (*Orlando Furioso*, xxxii), and that behind Panecia may lie the Fenicia whose story bears some resemblance to that of Ariodante and Genevra and was told by Bandello and retold by Belleforest.

Who wrote these plays there is no means of knowing. Two chance references suggest how active was the search for new productions. The one is in a fantastical letter of 1579 from Gabriel Harvey to Spenser, which if published, he feared, might cause him to be thrust upon the stage:

> to make tryall of my extemporall faculty, and to play Wylsons or Tarletons parte. I suppose thou wilt go nighe hande shortelye to sende my lorde of Lycesters, or my lorde of Warwickes, Vawsis, or my lord Ritches players, or sum other freshe starteupp comedanties unto me for sum newe devised interlude, or sum maltconceivid comedye fitt for the Theater, or sum other paintid stage whereat thou and thy lively copesmates in London maye laughe ther mouthes and bellyes full for pence or twoepence a peece.

And the other is in a letter of 1581 written from Sheffield by Thomas Bayly, an actor under the patronage of the Earl of

Shrewsbury, to Thomas Bawdewin of London, thanking him for a tragedy and begging him to procure:

librum aliquem brevem, novum, iucundum, venustum, lepidum, hillarem, scurrosum, nebulosum, rabulosum, et omnimodis carnificiis, latrociniis, et lenociniis refertum . . . qua in re dicunt quod Wilsonus quidam Leycestrii Comitis servus (fidibus pollens) multum vult et potest facere.

Must we suppose that in losing these plays we have suffered great loss? Certainly a historian of the drama must lament the loss of so many comedies, tragedies, and histories[1] written and performed immediately before the emergence of Peele, Marlowe, and their contemporaries. Yet the intrinsic merit of what has survived is slight. Where, in the few survivals from the professional players now to be considered, are

Le donne, i cavallier, l'arme, gli amori,
Le cortesie, l'audaci impresse

which are the main themes of Ariosto, Sidney, and Spenser? And if sometimes we catch glimpses of the themes, where are the handling suitable to these high matters and the answerable style? Perhaps we have been unlucky, yet if we are to believe Sidney all that we have lost are plays in which a young princess is got with child, delivered of a fair boy, who is lost, grows a man and ready to beget another child, 'and all this in two hours space'; or plays in which the scene changes from a garden to a rock to a cave, whence comes out a hideous monster belching fire and smoke; and so on. This section will end with some account of the only three extant plays of this period which can be assigned to the professional theatre, and if they are characteristic, we must acknowledge the justice of Sidney's compendious and contemptuous dismissal. Certainly they have none of the 'feeling of poetry', the 'poetical sinnewes', for which he was looking, and they are faulty in time and place.

The morality elements in *Common Conditions* and *Clyomon and Clamydes* have already been discussed, and something has been said of the deadly monotony of their fourteeners.[2] Both narrate 'nothing but the adventures of an amorous knight, passing from countrie to countrie for the love of his lady', a type which Gosson despised and attacked as 'meere trifles'. Both are romantic

[1] The tragedies and histories are treated below, pp. 146 f.
[2] Above, p. 70.

in the original sense of that word: the incidents are of the improbable kind found in some old romance. Indeed, they may be the first English plays in which romance is not swamped by native farce or pseudo-classicism or didacticism. The source of *Common Conditions*, the title of the edition of 1576 tells us, is 'the most famous historie of *Galiarbus* Duke of *Arabia*'. If this ever existed, it bore some relation in general to Greek romance and in particular to v. viii in Cinthio's *Hecatommithi*. The scene changes from Arabia to Phrygia, from a court to a wood to a ship. Lovers encounter with footpads and sea-pirates. One lover, like Shakespeare's Helena, suffers from unrequited love and is of low degree and the daughter of a physician; but whether she secures the lover she pursues is not known, as the end of the play is missing. So complicated are the adventures and separations that a brother falls in love with his sister and the sister takes service with her father, all unwittingly. The play is named after Common Conditions who is at once a Vice, a clown, a parasite, and the instrument of outrageous Fortune. The objection may be raised that the *Arcadia* contains incidents not less improbable; but like *The Faerie Queene* that romance *has* 'poetical sinnewes' and presents a view of life. It is difficult to imagine a play with less bark and steel for the mind than *Common Conditions*.

The attribution of *Clyomon and Clamydes* to Peele has been abandoned. Of about the same date as *Common Conditions* it was a very old-fashioned piece when it came from the press of Thomas Creede in 1599. The author is more a master of the rhetoric and metre at his disposal than is the author of *Common Conditions*, and behind the many adventures which take the spectator from Denmark to Suavia, to Macedonia, to the Isle of Strange Marches, to Norway, and involve the characters in shipwreck, combat with a flying serpent whose diet is womankind, imprisonment in an enchanter's fortress, there is a substantial background of chivalry, of tribute to 'a noble mind and eke a valiant hart'. Alexander the Great plays a part, but the plot has not been traced to 'the matter of Rome the Great', but to a French prose romance of chivalry, *Perceforest*, printed at Paris in 1528 in six folio volumes. In tight situations Rumour intervenes or Providence descends and ascends. Subtle Shift, the Vice, of whom there is no trace in the original, is hardly distinguishable from a cowardly comic and slightly knavish servant. Better low comedy—but hardly good enough to remind us of Launce—is supplied by a clownish shepherd and his dog. The play helps to make more intelligible Peele's handling of this type in *The Old Wives Tale*, but it does more. Here are most of the ingredients of romantic comedy: the characters whom Hamlet particularizes (the king, the adventurous knight, the lover, the lady, the humorous man, the clown); love at first sight; the courageous fidelity of the women, one of whom disguises herself as a page; a double ἀναγνώρισις; and at the end when 'the pageant is packed up, and all parties pleased', the satisfaction essential to romantic comedy when 'each Lord hath his Lady, and each Lady her love'.

Of a later date and utterly different in kind is *The Rare Triumphs of Love and Fortune . . . wherin are manye fine Conceites with great delight*, printed in 1589 and identified with 'A Historie of Love and ffortune' acted before the Queen at Windsor on 30 December 1582 by the Earl of Derby's players. The framework is mythological: Gosson would have condemned it out of hand for its representation of idols. Before an assembly of the gods Venus daughter of Jupiter and Fortune daughter of Pluto strive for chief authority over men and women. The ghosts of Troilus and Cressida, of Alexander, and Dido slain by Venus, of Pompey and Caesar slain by Fortune, and of Hero and Leander slain by both, appear in five dumb-shows accompanied by music. Jupiter breaks off the shows and orders the combatants to exercise their powers on flesh and blood. So the first act ends with the disputants taking their places, perhaps on the battlement of canvas provided by the Revels Office. At the end of the second and fourth acts of this 'pleasant Comedie' the 'triumph' (i.e. spectacle, pageant) is Fortune's, sounded with drums, trumpets, cornets and guns. The third act is Venus's, sounded with viols. In the fifth act Jupiter stops the contest, gods and men are reconciled, and the moral emerges that wisdom rules over both love and fortune. And so 'God save her Majestie that keepes us all in peace'.

In the play within the play the characters are the puppets of the two goddesses. Two lovers after much anguish come together, a knavish parasite and a clownish servant exhibit their humours, an old father goes mad when his books of magic are burned. The unknown dramatist has at his disposal fourteeners, poulter's measure, ten-syllabled couplets, blank verse, debate in alternate lines, and handles them with some competence. Here as in *Tamburlaine* Part I the only use of prose is to represent the incoherence of madness. The notable thing is that this courtly play was offered by a professional company of adult players. If more comedy of this period had survived from the professional theatre, the gap between Greene and the young Shakespeare and their predecessors might not seem so striking.

# VI

## TRAGEDY, *c.* 1540–*c.* 1584

THE earliest examples of regular tragedy written by natives of Britain were on biblical themes. Buchanan's *Jephthes* and *Baptistes* and Grimald's *Archipropheta* all belong to the fifteen-forties.[1] Grimald's play was the offshoot of the humanistic *tragicomoedia sacra*, and while the continental Buchanan was Senecan enough, he was writing before the English tragedians, learned and popular, became subservient to Seneca. The first known performance of a Senecan play in England was that of the *Troades* in Trinity College, Cambridge, in 1551–2, but it was during the decade 1559–69 that Seneca took the drama by storm. In 1559 appeared the first English translation, Jasper Heywood's *Troas*. Of far greater importance is the performance in 1562 of the first regular English tragedy, *Gorboduc*, followed after a short interval by two other Inns of Court plays in which the mark of Seneca is even more obvious, *Jocasta* and *Gismond of Salerne*.

*Enfin vint Sénèque*. The extent of his influence on English tragedy, academic and popular, would not have been so great if the themes, the doctrine, and the form had not proved congenial. The Elizabethans would enjoy the impression which his tragedies gave that crime meets its punishment in *this* life. They had the same appetite, or at least the same stomach, for sensational incident and violent passion, and most of them would have praised him, as did Cavalcanti, for disobeying the injunction to present no murders *coram populo*. Also they shared with him a taste for moral statement, for pithy *sententiae*, and a love of rhetoric. The son of a rhetor and bred in the schools of rhetoric, Seneca wrote for an age in which rhetoric was an absorbing passion and rhetorical amplification and embellishment with figures of words and sense were greatly valued and conspicuously employed.

His doctrine, it might be thought, would have repelled a Christian audience, but this was not so. The medieval *contemptus*

[1] Above, pp. 89 ff.

*mundi* had held that we are born in sin, linked to it before we are able to sin; that they are the happy ones who die before they are born, feeling death before they know what life is; that we come into the world weeping and we go hence weeping; that filthy in mind we are also filthy in body, breeders of vermin in life and in death meat for worms. This view of life presented in many medieval works written by good Christians, for example, the famous *De Contemptu Mundi* which Humphry Kerton translated into English in 1576 with no hint that its author was Pope Innocent III, is a view of life neither strained nor strange to any reader of Seneca's tragedies.

Again, it might be thought that a belief in the Roman goddess Fortuna was incompatible with the Christian belief, and indeed the medieval church and many a Protestant churchman of the sixteenth century held that it was, that Fortune was a mere figment of heathen invention which derogated blasphemously from the power and dignity of God. Yet while the goddess lost many worshippers in Christian times, the number of those who invoked her name did not decrease. She survived as a representation of the element of chance or what appears to be chance in human affairs. If one thought of Fortune as an agent permitted by God to interfere, apparently casually, in the affairs of man, it was possible to contemplate the co-existence of God and Fortune. She became a name for 'the inscrutable ways of God'. Others thought of her as an abstraction, as we do today, sometimes calling her Luck, and others, like the author of *Love and Fortune*, saw her uses as a poetic fiction.

Senecan tragedy abounds in references or invocations to Fortune. She is fickle and usually malignant. She rules the affairs of men without order and scatters her gifts with unseeing hand. While she abases the wicked she also despoils the good. (*Iniqua raro maximis virtutibus Fortuna parcit—Hercules Furens* 325–6.) Her favours are enigmatic and never more so than when bestowed upon the mighty. (*Ut praecipites regum casus Fortuna rotat—Agamemnon* 71–72.) She makes kings with mocking hand, and lifts up high only to bring down low. Therefore let man be modest in prosperity and tremble for the shifts of circumstance. They are the happy ones who, content with the common lot, with safe breeze hug the shore, and fearing to trust their skiffs to the wider sea, with unambitious oar keep close to land. (Cf. *Agamemnon* 101–7.) What remedy, then, for the mighty? For one thing keep a stiff upper lip. He that is willing to die is superior to Fortune. (*O quam miserum est nescire mori—Agamemnon* 611.) When adversity comes, stand firm with unfaltering foot. (*Haud est virile terga—Oedipus* 86.) While Fortune may take away wealth, it has no power over the mind. (*Fortuna opes auferre, non animum potest—Medea* 176.) These sentiments are not alien; they can be matched in scores of plays, Elizabethan and Jacobean. Seneca is not the only ancient author to give voice to them, but in him they are often condensed into a form so epigrammatic that the English were as incapable of translating as of forgetting them.

The vulnerability of the man of high estate is the central theme of Elizabethan tragedy as of medieval. In the Middle Ages tragedy was a matter for narrative verse and prose, and the definitions had not yet confined it to drama. Two of the most famous are Dante's and Chaucer's. In a Latin letter to Can Grande Dante explains that he has called his poem a comedy because it began in misery (*Inferno*) and ended in happiness (*Paradiso*), whereas tragedy is in the beginning good to look upon and quiet, but in the end foul and horrible. And similarly in the Prologue of 'The Monk's Tale' Chaucer writes of tragedy being a certain story

> Of hym that stood in greet prosperitee
> And is yfallen out of heigh degree
> Into myserie, and endeth wrecchedly.

(Thomas Heywood in his *Apology for Actors*, 1612, provides a vulgarization: 'Comedies begin in trouble, and end in peace; tragedies begin in calms, and end in tempest.') The two best English biographies of the sixteenth century, Roper's *More* and Cavendish's *Wolsey*, are both tragedies which narrate the fall of a man 'out of heigh degree'. In Roper Fortune plays no part at all. She has no hold upon a man so indifferent to wealth and to honours; and no blind chance, only the integrity of a saint, leads More to martyrdom. But Cavendish's life is punctuated with references to false and fickle Fortune who is then most deceitful when her servant is in highest authority.

> O madness! O foolish desire! O fond hope! O greedy desire of vain honours, dignities, and riches! Oh what inconstant trust and assurance is in rolling fortune!

The most striking examples in the Elizabethan age of men who have fallen from high estate are in the collection of verse complaints entitled *A Mirror for Magistrates*. The suggestion for the book, as for 'The Monk's Tale', came from Boccaccio's *De Casibus Virorum Illustrium*, a work which Lydgate versified from a French version in the fourteen-thirties. Boccaccio's moral for the most part is that of ascetic *De Contemptu*, but there are exceptions, as Professor Willard Farnham has observed in an admirable book. No exception is more striking than the sympathetic comment on the fate of Pompey, a comment that approaches the true nature of tragedy as interpreted by Aristotle. 'If such greatness suffered a fall, what may we suppose might happen to us? We ought certainly to pity Pompey; but we ought much more certainly to fear for ourselves.' But only momentarily does he see tragedy in the grand manner, see that 'tragic event is the product both of fate for which the individual is not responsible and of characteristic deed for which he is'.

The English counterpart of *De Casibus*, *A Mirror for Magistrates*,[1] first published in 1559, contained nineteen tragedies: seven were added in 1563, Sackville's 'Buckingham' among them together with his Induction, and others in 1578 and 1587. And there were sequels of which the last appeared in 1610. The twofold nature of tragedy in these collections of 'Complaints' is shown by the words which appear on the titles of 1559 and 1563: 'Wherein may be seen by example of other, with howe grevous plages vices are punished: and howe frayle and unstable worldly prosperitie is founde, even of those whom Fortune seemeth most highly to favour.' (To George Puttenham a little later the fall of princes shows 'the mutabilitie of fortune, and the just punishment of God in revenge of a vicious and evill life'.) The editor, William Baldwin,[2] explains in his dedication to the nobility and lesser 'magistrates', that the aim is to show as in a looking-glass that Justice is the chief virtue and the administration of it the 'chiefest office': to show also that while God suffers many offices to be occupied with unjust

[1] See C. S. Lewis, OHEL.

[2] He was also the compiler of a *Treatise of Moral Philosophy* or collection of sayings of the wise printed in 1547 and often later. Of his minor works the most interesting is the lively prose satire *Beware the Cat* (entered 1568–9 but written earlier). A play *Love and Live* is lost: it was a freakish morality with 62 characters the names of which all began with L.

rulers, 'yet suffreth he them not to skape unpunished, because they dishonour him'.

Only a few of the twenty-six men and one woman (Jane Shore) who speak their complaints in the first two editions are, like Henry VI, innocent. Most are rulers of varying degrees of guilt whom God does not suffer to escape unpunished. Yet they cry out more upon Fortune than upon God. Fortune allows their vices free scope till they rise to the top of her wheel, then hurls them down when they are most sure. A few challenge the sway of Fortune. Buckingham blames not Fortune, not the fates, not Jove, but 'the fyckle fayth of commontye alone'. Mowbray is Senecan and medieval in maintaining that Fortune controls material things, but 'hampreth not the harte'. Cade, 'one of Fortunes whelpes', is sure of this one point that, whatever the influence of the planets or of our 'complexions', 'Our lust and wils our evils chefely warke'. And Henry VI in seeking the cause of man's 'heavy happes' dismisses astronomy, the humours, chance, for two chief causes:

> The chiefe the wil divine, called destiny and fate,
> The other sinne, through humours holpe, which god doth highly hate.

Here and there the conditions which make great tragedy possible suggest themselves as if waiting only for some spark of genius to ignite and unite them.

The year of the first edition of Baldwin's *Mirror* was also the year of the first of many English translations of Seneca. Jasper Heywood of Merton and All Souls[1] led the way with his *Troas* (1559), *Thyestes* (1560), and *Hercules Furens* (1561), and Cambridge soon followed with Alexander Neville's *Oedipus* (1563), John Studley's *Agamemnon* and *Medea*, and Thomas Nuce's *Octavia* (all 1566). Studley's *Hippolytus* and *Hercules Oetaeus* together with Thomas Newton's *Thebais* appeared first in the collected *Ten Tragedies* edited by Newton in 1581. The translators were young men with their way yet to make in the world—these were their first fruits—and they drew

[1] 1535–98. The younger son of John Heywood and the uncle of John Donne. He was for three years Fellow of Merton—the last Merton 'King of Beans'—and moved to All Souls in 1558. A few years later he left England for religion's sake, entered the Society of Jesus, and joined the English mission. In this hazardous occupation he showed great courage but little discretion. Discovered in 1583 he was exiled in 1585 and died in Naples.

attention to themselves by dedicating their work to a famous statesman or to the Chancellor of their university, even to the Queen. Either they admit that they are offering a little toy but have every intention of benefiting their country with weightier matters or they defend their labours by pointing to the salutary lessons which the tragedies teach. Both Studley and Neville shelter behind Erasmus's praise of Seneca as a grave, virtuous, and Christian ethnic, praise which Erasmus did not direct particularly at the plays. Seneca, though heathen, may teach men to embrace virtue and shun vice. He admonishes all men of their fickle state, declares the inconstancy of wavering fortune, expresses in the liveliest fashion the just revenge and fearful punishments which await the horrible crimes with which the wretched world abounds. Far from countenancing ambition, tyranny and incontinency, no one among the heathens

> with more gravity of Philosophicall sentences, more waightynes of sappy words, or greater authority of sound matter beateth down sinne, loose lyfe, dissolute dealinge, and unbrydled sensuality.
>
> [Newton, 1581.]

Of the merits of his own translation each man professes to think little, but he can be unsparing in his praise of another's. Heywood and Neville, says Studley, write with such excellence 'that in reading of them it semeth to me no translation, but even Seneca hym selfe to speake in englysh'. More often they point out the impossibility of preserving Seneca's peerless sublimity and royalty of speech in a language so corrupt, so gross, and so barbarous as English. Heywood and Studley choose fourteeners, but fourteeners with a caesura so emphatic and so fixed that in the early (octavo) editions they were printed as eights and sixes. A shorter line is often chosen for the choruses. Whatever the length, the lines are unspeakable upon a stage, but then Neville is the only translator who conceives of the possibility that his version might be acted. How far they introduced a knowledge of Seneca to readers who had no Latin we have not even the right to guess. Certainly there was nothing here from which the authors of *Gorboduc* might profit. *Troas* is the only play twice printed in octavo and there was no call for a reprint of the collected edition. The merits of the translators are slight. Heywood is better in the lyric measures of the choruses, and Studley has vigour, sometimes too much vigour, as when he turns 'ardet felle siccato iecur' (in description of Hercules in torment) into 'And from my frying Ribs (alas) my Lyver quite is rent.' Is this what William Webbe wanted when he said that 'to the Tragical wryters belong properly the bygge and boysterous wordes'? The additions are more interesting than the excisions. In *Troas* Heywood brings into the play the ghost of Achilles thirsty for 'vengeance and bloud', of whom in Seneca we know only by report, and augments his author by ninety-two lines:

> The soile doth shake to beare my heavy foote
> And fearth agayn the sceptours of my hand,
> The poales with stroke of thunderclap ring out,
> The doubtful starres amid their course do stand,
> And fearful Phebus hides his blasing brand.
> The trembling lakes agaynst their course do flyte,
> For dreade and terrur of Achilles spryte.

Here, Professor Charlton says, 'fully equipped in manner, speech, and dramatic function, the Senecan ghost first enters Elizabethan drama—and he is by law, at all events, Heywood's ghost, not Seneca's at all!' When these translators add, they attempt to 'overgo' Seneca in horror and bombast.

We may turn to better work in better metre. Baldwin's *Mirror*, like Shakespeare in all his Histories but one, chose its themes from the period Richard II to Henry VIII. Not until 1574, in John Higgins's sequel, were the chronicle histories searched for British examples. Long before Higgins, however, two dramatists, one of them a contributor to the *Mirror*, wrote an actable play on a theme from British history which has every right to be called the first regular English tragedy in dramatic form. *Gorboduc* or *The Tragedy of Ferrex and Porrex* was acted by the gentlemen of the Inner Temple at their Christmas revels in 1561–2 and a few days later at Whitehall before the Queen on 18 January 1562. This is the only known example of a play presented at Elizabeth's court between Twelfth Night (6 January) and Candlemas (2 February); the tribute was well deserved. The joint-authors, Thomas Norton and Thomas Sackville,[1] both of the Inner Temple, were neither of them

[1] Thomas Norton (1532–84), son of a wealthy citizen of London, joined the Inner Temple in 1555 already a zealous reformer. He married one of Cranmer's daughters. To about the same date as *Gorboduc* belong the metrical psalms he contributed to 'Sternhold and Hopkins'. A successful lawyer, he became standing

'punies of the law'. Norton was twenty-nine and Sackville twenty-five, and both had sat in Elizabeth's first parliament of 1559. Norton's translation of Calvin's *Institutes* had been published in 1561, the first of many works by this zealous Protestant and industrious lawyer and parliamentarian. Sackville may already have written the Induction to the *Mirror* and the Complaint of the Duke of Buckingham.

Their plot is derived ultimately if not directly from Geoffrey of Monmouth, perhaps the only begetter of the long line of British kings descended from Brute, great grandson of Aeneas. Of these Gorboduc was the eighteenth ruler, as Brute was the first and Lear the tenth, and with him and his two sons the line of Brute came to an end, no rightful successor being left alive to succeed them. The plot is sensational. Like Lear, Gorboduc in his old age divides his kingdom between his children. Egged on by flattering parasites, Ferrex and his younger brother Porrex go to war, Ferrex is slain; in revenge for the murder of her favourite son Queen Videna kills Porrex, the people rise and kill both King and Queen, and as the play ends the nobles are at loggerheads and the country is in ruins.

Sidney with his '*Gorboduc* . . . is full of stately speeches and well sounding Phrases, clyming to the height of *Seneca* his stile' was perhaps responsible for the view that the play is overwhelmingly Senecan. True, we are reminded of Seneca by the bloody sensationalism of the plot, by the division into acts which gave the play form and dimension, by the chorus after each of the first four acts, by the length of the speeches, by the use of messengers and the banishment of all violent action from the stage, by the sententiousness of some of the speeches. But not all these characteristics are peculiarly Senecan, and there is much in the play that is in no way Senecan: the dumb-shows the origin of which is obscure but which may owe something to the *intermedii* of Italian drama and more to the use of visual allegory and symbol in English masque and pageant; the insistence that every man invites by his own

counsel to the Stationers' Company in 1562 and Remembrancer to the City of London in 1570.

Thomas Sackville (1536–1608), whose father was a court official (Sir Richard 'Fillsack'), joined the Inner Temple in 1555 after a brief sojourn at Oxford. His rise to power and political fame dates from the death of his father (1566) from whom he inherited a vast fortune. He won the favour of the Queen—to whom he was distantly related—and her successor, became Lord Buckhurst in 1567, and died Earl of Dorset.

actions the retribution of the gods, a view that exists side by side with the more Senecan doctrine of an overhanging Fate and of a curse that may threaten with destruction a whole race; the political moral; the strong spirit of patriotism and love for 'the common mother of us all, Our native land, our countrey'. These are native English, and so too are the counterbalancing good and evil counsellors who seem to derive from the morality play. So too is the style, and if we look for evidence of the authors' reading we find little borrowing from Seneca and much from elsewhere. For example, 'Blood asketh blood' (IV, Chorus), to appear so memorably in *Macbeth*, derives from *Genesis* ix. 6; the *sententia* 'Oh no man happie, till his ende be seene' (III. i. 11) is Solon's answer to Croesus and had already been given English currency by 1545; and the vivid

> Then saw I how he smiled with slaying knife
> Wrapped under cloke

would not have been so vivid but for Chaucer's more vivid 'The smyler with the knyf under the cloke.'

Sidney censures *Gorboduc* for not observing the unities of time and place, unities first codified in all their harsh rigour by Castelvetro eight years after the play's production. He does not censure it for failure to observe the unity of action, yet here is one of the most interesting of the play's departures from Seneca, and it is a departure dictated to the authors by their main purpose. Their play is not so much the tragedy of Gorboduc, his wife, and their sons, as the tragedy of a divided state. Gorboduc and his race are extinguished after the fourth act, and the last is given to the state of Britain. In this respect the play is allied not to Seneca but to *Respublica*, to the chronicle histories yet to be written, and to *A Mirror for Magistrates*. There is the initial error of dividing the kingdom in the lifetime of the king, for as the first dumb-show signifies with faggot and sticks 'a state knit in unitie doth continue strong against all force. But being divided, is easely destroyed'. The king's intention is debated in the second scene by three counsellors of state, and is opposed only by the wise counsellor, Eubulus, with whose sentiments the authors identified themselves. The evils that he foresees come to pass: 'Divided reignes do make divided hartes.' It is Eubulus too who condemns the murder of Gorboduc, and the sentiment

Though kinges forget to governe as they ought,
Yet subjectes must obey as they are bounde

is one we have met before and shall meet again.[1] It is he who advocates the speedy suppression of the revolt by the common people, 'more wavering than the sea'. It is he who when the news comes that an ambitious noble seeks the crown for himself speaks the lines which recall the greater lines of the Bishop of Carlisle in *Richard II*:

And thou, O Brittaine, whilome in renowme,
Whilome in wealth and fame, shalt thus be torne,
Dismembred thus, and thus be rent in twaine,
Thus wasted and defaced, spoyled and destroyed,
These be the fruites your civil warres will bring.

And lastly it is he who while prophesying years of civil war sounds a note of hope for the days he will not live to see:

Of justice, yet must God in fine restore
This noble crowne unto the lawfull heire:
For right will alwayes live, and rise at length,
But wrong can never take deepe roote to last.

The play was first printed in 1565 in an edition which the second edition of *c.* 1570 called unauthorized. The text of 1565, however, is good, and is the only authority for the statement that Norton wrote the first three acts and Sackville the last two. Some recent critics have been inclined to accept this division with the modification that Sackville wrote the first scene of the play and Norton the last. Certainly the opening lines, conventional in all but their musical phrasing, remind us of the poet of the 'Induction'. But attempts to distinguish by tests of metrical analysis and vocabulary have not proved more conclusive than the attempts to detect divergent political views. The chief message of the play in its bearing on contemporary politics is that in the absence of an heir to the throne a parliament should be called to establish the succession in the lifetime of Elizabeth, for only so can her name and power make a parliament of force 'By lawfull sommons and authoritie.' The message might have been subscribed not only by Sackville and Norton but by Burghley and thousands more of her loyal subjects, fearful of an interregnum and the sedition and anarchy which might ensue.

[1] Above, p. 37.

The style throughout the play is remarkably consistent and of a uniform gravity. Of the archaisms of the 'Induction' there is no trace, and as unsuitable to the heroic style would have been the colloquial. The authors keep decorum. In the language there is little that is obsolete, and the rhetoric shows none of the over-indulgence in tropes and figures which makes ridiculous to modern taste so much of the verse and prose of that age. The classical *exempla* are infrequent, and are there to illustrate not merely to adorn. Pope praised the play for its lack of bombast and affectation, 'the two great sins of our oldest tragic writers', and he also commended 'an easy Flow in the numbers'. For the verse does flow. While the voice must pause a little at the end of each line, the sense flows on sometimes for a long paragraph. A speaker can get more variety out of the verse than may be supposed, whether in the scenes of debate or those of passion. To most modern critics, however, the play is dull. They deplore the absence of any action, miss 'the sprightly animation of reciprocal dialogue', and find the long political speeches tedious. Yet it is a dignified historical tragedy, the first of its kind in English, written by two young men of unusual ability, and we may be grateful that dramatic blank verse got off to so good a start.

The merits of *Gorboduc* appear the more striking if we compare the play with three later Inns of Court tragedies, *Jocasta*, *Gismond of Salerne*, and *The Misfortunes of Arthur*, all following the five-act structure. The authors of *Jocasta*—George Gascoigne (Acts II, III, and V) and Francis Kinwellmersh (I and IV)—announce that their play was 'translated and digested into Acte' from Euripides, but in fact they translated an Italian adaptation of the *Phoenissae* by Lodovico Dolce. As Dolce's *Giocasta* is based on a Latin translation, the English play is three removes from the Greek. Where Dolce omits, he omits passages and allusions unintelligible to a sixteenth-century audience: where he adds, and he adds freely, he adds moralizing passages often suggested by his favourite author Seneca. For this reason his play was much more congenial to his listeners. It suited them also that the play is full of reversal of fortune, for even if violent action was excluded from the stage there was an advantage in having plenty of things happen. Inspired no doubt by *Gorboduc* the Englishmen add dumb-shows, so supplying a measure of spectacle otherwise denied to them; and they follow the same model in the use of blank verse, though they prefer rhyme royal for the choruses. Yet whereas the main emphasis in *Gorboduc* is topical and political, here it is moral, and the morality is characteristically Elizabethan, a medley of Greek, Latin, Italian, and English moral reflexion. Here are themes and attitudes which re-echo through Elizabethan and early Stuart drama. Life is tedious and short and full of pain, and the sooner a man is rid of it the better; it is never firm, never staid, but whirls about with the wheels of restless time. The heavens (or Fortune) rule the rolling life of man, and it is vanity to strive against fate or necessity; all men are tied to Fortune's wheel. At the same time content is possible if only we can abide with patient hearts what is sent to us. Unbridled ambition brings ruin in its train, whereas wise men sustain with patience and a quiet mind the blows of slippery Fortune. In the words of Oedipus (V.v.):

> Let fortune take from me these worldly giftes,
> She can not conquere this courageous heart, . . .
> Do what thou canst I will be *Oedipus*.[1]

The morality the Englishmen rub in with marginal comment. 'The courte lively painted' faces a passage on the vanity of court-life; wise counsel given to Antigone is called 'A glasse for yong women' and her kindness to Oedipus 'The duty of a childe truly perfourmed'; and the last speeches of her father are glossed 'A Glasse for brittel Beutie and for lusty limmes' and 'A mirrour for Magistrates'. In the epilogue added by Christopher Yelverton the moral becomes more political. The deaths of Eteocles and Polynices are 'the fruit of high-aspiring minde'.

> The golden meane, the happie doth suffise,
> They leade the posting day in rare delight.

*Jocasta* was acted at Gray's Inn in 1566, and *Gismond of Salerne* by the Gentlemen of the Inner Temple before the Queen at Greenwich perhaps in the same year, perhaps in 1568. Two manuscripts and a fragment preserve this play as first written, and a quarto of 1591 (*Tancred and Gismund*) presents the play 'Newly revived and polished according to the decorum of these daies.' Among the changes are the substitution of blank verse for rhymed and the addition of dumb

[1] Dolce: '*Fa quel che puio; io farò sempre Edipo*'.

shows. The polisher was Robert Wilmot, author of Act V and the Epilogue of the original version: the four other collaborators are all unknown to fame except Christopher Hatton who wrote Act IV. In old version or new the play is not much more than a Senecan-Cinthian curiosity, yet the tale which gave these 'young heads' of the Inns of Court their plot gives promise of romantic tragedy. They take the story not from Painter but direct from the *Decameron* (IV. i). It is the story of how a tyrannical father Tancred forbids his widowed daughter Gismond to remarry, surprises her with her lover, in revenge sends her lover's heart to her in a cup, a cup from which she drinks (*coram populo*) after adding poison. The attraction of the theme to any Senecan-minded dramatist is obvious. In contrast to the three other Inns of Court tragedies of this date the play is wholly concerned with 'the matter of Loue', a matter, as Wilmot sagely observes, 'as old as the world'. But the characters appear as the puppets of the gods, Cupid presiding over the passions of the lovers and Megaera over the cruelties of the father, the lovers never appear on the stage together, the speeches are long and set and often in soliloquy, the choruses tediously moral. In 1591 Wilmot, then in orders, made much of the moral, yet to the modern mind the moral of these revolting horrors must remain doubtful. (Some of the moral reflexion is from Seneca and some—including Prologue Cupid—from Dolce's *Didone*.) The moral Wilmot reads into the play is not more forced than that Arthur Broke reads into an early play on Romeo and Juliet, the earliest romantic tragedy in English, of which we hear, to be based on an Italian *novella*. In the address to his poem *The Tragical History of Romeus and Juliet* (1562) he tells us he had seen the same argument 'lately set forth on stage' and set forth well. This 'history' of two star-crossed lovers is to us the very type or figure of romantic tragedy, yet the lost play and Broke's poem alike preached the dangers of neglecting the advice of parents and following that of superstitious friars and of using the honourable name of marriage to cloak the shame of stolen contracts.

*The Misfortunes of Arthur* cannot have for us the interest of *Gorboduc*, *Jocasta*, and *Gismond*, for it came too late to influence the popular stage: indeed it may well have been influenced by it. *Tamburlaine* had taken the London stage by storm a year before 28 February 1588 when the *Misfortunes* was acted before the Queen at Greenwich by the Gentlemen of Gray's Inn; and *The Spanish Tragedy* may also have been earlier. The play was seen through the press by Thomas Hughes[1] before 25 March, for the quarto is dated 1587. Blank verse with stanzaic forms for the choruses was by now standard form for academic drama, and so were dumb shows, two of them supplied for this play by Francis Bacon and Christopher Yelverton. The play is thoroughly Senecan, and much of it mere translation especially from *Thyestes*. The story came from the 'bold bawdry' of Geoffrey of Monmouth (with touches from Malory): how Uther Pendragon begat by Igerna, wife of Gorlois, the twins Arthur and Anne, how Arthur begat Mordred by his sister, how Mordred committed adultery with Arthur's wife Guenevora, and of the battle between Arthur and Mordred which extinguished the line of Pendragon. Here appear two Senecan traits, the ghost and stichomythia, which we miss in *Gorboduc*, *Jocasta*, and *Gismond*: but they may have already appeared in Kyd. The Ghost is Gorlois, spirit of revenge, who is derived from Seneca's Tantalus. An example of Seneca's pervasive influence is found in

> Him, whom the Morning found both stout and strong,
> The Evening left all groveling on the ground (Epilogus 40–41)

which is Seneca's

> Quem dies vidit veniens superbum,
> hunc dies vidit fugiens iacentem (*Thyestes* 613–14)

and appears at the close of *Sejanus* as

> For, whom the morning saw so great, and high,
> Thus low, and little, 'fore the 'even doth lie.

And yet, as recent critics have observed, the play is not solely Senecan, but in its hatred of tyranny, its detestation of the crime of usurpation, its condemnation of political ambition, it is adapted to the Elizabethan scene. Un-Senecan too is the concern with the welfare of the state:

> Thou Realme which ay I reverence as my Saint,
> Thou stately Brytaine th'auncient type of Troy.

And most Elizabethan is the prophecy of the Ghost of Gorlois before he descends to hell that many ages hence a virtuous

[1] He matriculated from Queens' College, Cambridge, in 1571 and became Fellow in 1576.

virgin, 'braunch of Brute', 'Shall of all warres compound eternall peace.'

*Gorboduc*, *Jocasta*, *Gismond*, these are the more important in these formative years before Kyd and Marlowe because they were shaped for the stage and in English. Undramatic as the models they follow may be, these authors write actable plays, and although they write coterie drama, the coterie was wide enough to include the court. Presumably the lawyers presented their plays in English for the reason that they depended in their profession on the rhetorical and forensic use of their own language. At the universities Latin was the medium for tragedy and usually for comedy. Cambridge in these years was busying itself with comedy[1] and in tragedy offers nothing worthy of mention before Dr. Thomas Legge's *Richardus Tertius* (acted at St. John's College, Cambridge, in 1579). This is of great interest as being—if we except Bale's *King John*—the first play based on English history. It antedates by perhaps a decade the appearance of such a theme on the popular stage. We find in turning to Oxford that until just before the successes of Kyd and Marlowe on the popular stage it has as little to offer in tragedy as Cambridge. If we pass over Canon James Calfhill's lost *Progne* (acted at Christ Church in 1564), chiefly notable because that scourge of the academic stage John Rainolds (then an undergraduate) took the part of Hippolyta, we find nothing to our purpose before William Gager of Christ Church.[2] In his day a notable figure in the dramatic life of Christ Church and his university, his occasional verses and his plays earned him a reputation for elegant Latinity. If instead of writing in Latin he had contributed in metre and rhythm, in diction and phrasing, to his own language, he would be better remembered; but to wish that he had done so would be to infuriate those who still admire his skill in a learned tongue. He was the author of three tragedies: *Meleager*, acted 1582, printed 1592; *Dido*, acted 1583; and *Ulysses Redux*, acted and printed 1592, a play with a happy ending which he calls a *Tragoedia Nova* because almost to the end the spectators hover in suspense. His one comedy *Rivales* (acted 1583) is

[1] Above, pp. 109 f., 113 f., 116.

[2] William Gager (1555/60–1622) entered Christ Church from Westminster in 1574, and took his M.A. in 1580, his D.C.L. in 1589, the year he left Oxford to become Chancellor of the diocese of Ely.

unfortunately lost: on the strength of it Francis Meres styled him one of 'the best for Comedy amongst us'.

When he began to write, the work of the neo-Senecan Robert Garnier, whose seven tragedies and one tragi-comedy are the most substantial body of dramatic work to come from sixteenth-century France, was almost completed: Garnier's first play *Porcie* and his last *Les Juifves* belong respectively to 1568 and 1583. Gager appears not to have noticed them, and not until 1592 and 1594, when the Countess of Pembroke published her translation of *Marc Antoine* and Kyd his translation of *Cornélie*, did Garnier's work come to England. When it came, it was to inspire, under the patronage of Sidney's sister, a school of closet drama of which the shining examples are Daniel's *Cleopatra* (1594) and *Philotas* (1605) and Fulke Greville's *Alaham* (written *c.* 1600) and *Mustapha* (1609). But these belong more to the history of poetry or ideas than of drama.

Garnier and Gager are proud to be Senecans, but both depart from their master and do so in different ways. While Garnier's plays are not closet drama in the extreme sense that *Cleopatra* and *Mustapha* are, they were not intended for the stage and are devoid of stage directions: Gager's plays were sumptuously produced in his own college, sometimes more than once, and contain elaborate stage directions. Indeed, the scenic effects in his friend George Peele's production of *Dido* were so remarkable that Holinshed (or rather Fleming) dilated upon them, although there is nothing in them that would have daunted the mechanics responsible for the court and street pageantry of earlier centuries. It will be noticed that the cry of hounds which went down so well at Christ Church in 1566 was repeated in 1583.

> The queenes banket (with Eneas narration of the destruction of Troie) was livelie described in a marchpaine patterne, there was also a goodlie sight of hunters with full crie of a kennell of hounds, Mercurie and Iris descending and ascending from and to a high place, the tempest wherein it hailed small confects, rained rose-water, and snew an artificiall kind of snow, all strange, marvellous, and abundant.

In other ways Gager was lavish in comparison with Garnier. Whereas the Frenchman limits himself to a very few characters, dispenses almost entirely with physical action, is not enamoured of horror, expresses his theme mainly through monologue and chorus, represents his history (as Jean de la Taille advised in 1572) 'en vn mesme iour, en vn mesme temps, et en vn mesme lieu', the Englishman crowds his scene with characters, is indifferent to the unities of time and place, like Seneca is not afraid of *atrocitas* and presents it upon the stage, breaks up monologue by a plentiful use of the confidant, has an eye for situation and the excitement and bustle and suspense which may be squeezed from it, and especially in *Ulysses Redux* (where he takes form from Seneca but his matter, *Lignumque, caementumque, lapidesque optimos,* from Homer) rejects elevation of language for the sake of the vivid and the familiar, and defends the mingling of comedy with tragedy. In short, he might have become a popular dramatist if he had wished. *Effudi potius quam scripsi*, he declared, and in the same sentence chose to be weighed in the balance of popular judgement. In one respect, however, Garnier's plays conform more than Gager's to *Gorboduc* or *The Misfortunes of Arthur*: as one of his critics has remarked, he never condescends 'to the vulgarity of mentioning France', but he wrote at a time of civil war, and the reader is never in doubt that be the scene Egypt or Troy or Jerusalem the woes that are being lamented are those of his own country.

It is not easy to estimate the effect which Seneca had upon English tragedy in these early years. When an author crosses the Channel he always suffers a sea-change, and the more so if he is so congenial that his characteristics fuse with native elements. A taste for horror, a taste for rhetoric, a taste for ethical commonplace, these would have marked English tragedy if Seneca had remained unknown. A later generation of dramatists, it has been urged, the generation of Jonson, Chapman, Marston, absorbed more of Seneca and to better purpose than did these pioneers. Yet the reverence with which the Greekless Elizabethans regarded Seneca the tragedian impelled the dramatists of the universities and the Inns of Court to emulate him and in so doing to set an example of serious drama at a time when popular drama in all except comedy was almost totally devoid of art. These academic writers adopted the external form of act and scene. Possibly this might have come to tragedy by way of Plautus and Terence: in fact it came by way of Seneca. It is a form which may induce, though it cannot compel, a dramatist to present a sequence of cause and effect and to reduce complexity to unity. But nothing is more important in these years, we may say with the hindsight of the historian, than the invention of blank verse and its transference to drama in *Gorboduc* and *Jocasta* ready for the hands of Kyd and Marlowe, an invention which was to rid the drama of the monstrous monotony of the fourteener.

Popular dramatists were slow to follow the example of *Gorboduc* in this respect if we may judge from the extant remains. R. B.'s *Appius and Virginia* (entered 1567), John Pickering's *Horestes* (printed 1567 and probably to be identified with the *Orestes* acted at court in 1567–8), Thomas Preston's *Cambyses* (entered 1569 and possibly acted at court in 1560–1) all give the *dramatis personae* on their titles while the two last divide up the parts for six and eight players respectively. They are not divided into acts and scenes. The identification of R. B. with Richard Bower, master of the children of the Chapel from 1561 to 1566, is doubtful: *Appius and Virginia* has plenty of song, though not always of the kind we should think suitable for children, but song is not confined to children's plays. In the play as printed there is no hint of a court performance. Thomas Preston of Eton and King's won the Queen's favour in 1564 by disputing with Cartwright and playing in Haliwell's *Dido*, and in 1589 rose to be vice-chancellor, but some find it incredible that such a man could have written so crude a piece as *Cambyses*. Perhaps the author was the Thomas Preston of whom nothing is known except that he wrote several topical ballads, among them 'A geliflower of swete marygolde, wherein the frutes of tyranny you may beholde', the very theme of *Cambyses*. John Pickering too is a mere *nominis umbra* unless we identify him, as there is good evidence for doing, with Sir John Puckering.[1]

All three have morality elements, elaborate the comic scenes in tumbling verse, and use the fourteener for serious verse. The monotony of their verse and the crudity of their rhetoric have already been condemned.[2] In *Appius and Virginia* the morality

[1] 'Pickering' is a variant form of 'Puckering'. Sir John Puckering (1544–96) was admitted to Lincoln's Inn in 1559 and called to the Bar in January 1567. *Horestes* is full of legal arguments, and in 1566 Lincoln's Inn came into prominence as a centre of anti-Marian feeling. Puckering rose to be Lord Keeper and became a leading enemy of the Queen of Scots and advocate for her execution. In 1567 he was twenty-three years of age, two years younger than Sackville of the Inner Temple when *Gorboduc* was acted.

[2] Above, pp. 70 ff.

element is strong not only in the comic scenes between the Vice and his associates but in those where Conscience disputes with him for the soul of the unjust judge, Appius, where Comfort comes to the assistance of the sorrowing father Virginius, and Justice and Reward distribute praise and penalty. In *Cambyses* allegorical and even mythological characters appear, but the comic scenes are more detached from the serious than in *Appius and Virginia*: the King needs no Vice to spur him on in cruelty and tyranny. The opening scenes show him as a just monarch punishing a corrupt judge, the other scenes display instance after instance of brutality and at the end retribution. If these serious scenes could be taken seriously, the comic scenes would provide a ghastly mirth. Preston's probable source has been shown to be Taverner in his *Garden of Wisdom* (1539, etc.) who derives from the German Johan Carion who derives from Herodotus. In source and play the moral is the evil of tyranny and the certainty of divine retribution. In one respect and only one does *Cambyses* fulfil the right use of 'high and excellent Tragedy' as defined by Sidney: it 'maketh Kinges feare to be Tyrants and Tyrants manifest their tirannical humors'. But before it could have stirred in Sidney or Shakespeare 'the affects of admiration and commiseration' it would have had to be in another and a better 'vein'.

*Horestes* (i.e. Orestes) is the most interesting play of the three, though interesting more for what it promises than for what it does. Whether considered as history, tragedy, or that special form of the Elizabethan tragedy, the revenge play, it repays attention. A play 'Orestes' was acted at court at the Christmas or Shrovetide festivities of 1567–8. To say that *Horestes* is too crude a production for court performance is to underestimate the play and to overestimate the standards of court taste at that time. It demands some scenic effects and especially a castle gate with battlements above. It was most unusual, however, if not unprecedented, to act at court a play which had already been printed; the play may have been published just before 24 March 1567–8 and almost immediately after performance. Though crude, it is without the gross bombast and absurd ineptitudes of *Appius and Virginia* and *Cambyses*, the songs are spirited, and there is little that does not advance the action. A defence might even be put up for the comic scenes on the ground that they give at a realistic level the common people's attitude to war. The Vice is *sui generis*. In so far as he is the old comic Vice he speaks in tumbling verse; in so far as he is Revenge and a messenger of the gods he speaks in fourteeners. The decision of Horestes, whether he should or should not revenge the murder of his father on his own mother, is not made to appear easy. Nature counsels him for the sake of compassion and mother-love to spare her: Revenge and the friendly Idomeus (Idomeneus) urge revenge. When Clytemnestra is captured, she is put to death not by her son but by Revenge, and Revenge is banished by the sorrowing king. Before Menelaus, seeking revenge for the killing of his sister, Horestes pleads that he obeyed the gods' command, and is exonerated. 'I dyd but that I could not chuse.' And the chain of crime is broken by the marriage of Horestes to Menelaus's daughter. The duty of revenge or punishment is stated—of public revenge, of course, not private. Vice and virtue flow from the palace of a king as water from a fountain, and therefore revenge for a deed so evil is necessary to teach the infected to subdue their will and to prevent others from becoming infected by ill example. The play ends suitably with a dialogue between Truth and Duty in which the curse of dissension and the blessing of unity are stated and prayer made for Elizabeth and her counsellors and commonalty, not forgetting 'my Lord Mayre, lyfetennaunt of this noble Cytie'.

That there is general political reference is obvious: some have found in the play a particular application to Mary Queen of Scots. In 1567 Darnley was murdered, Mary married Bothwell (whom many supposed to be the murderer), Bothwell fled his country, and Mary was imprisoned and forced to abdicate. Attempts to allegorize the play and make Agamemnon stand for Darnley, Egisthus for Bothwell, and the Queen for Clytemnestra are bound to fail; no one has supposed that the one-year-old James VI can be equated with Horestes. But the argument that the play is a deliberate presentation of the circumstances in which an anointed queen may justifiably be dethroned and punished is cogent. The Protestants of England, writes Professor James Phillips, 'approved the practice but deplored the principle of dethroning and imprisoning a sovereign queen'. Mary and Clytemnestra—the two were linked by contemporary satirists—are special cases where punishment is demanded by the laws of God, Nature, and Nations. The topical reference and to some extent the attribution to Puckering are supported by the fact that the legal and philosophical arguments justifying Horestes's vengeance are all additions to the story told in Caxton's *Recuyell of the Histories of Troy*, the dramatist's source.

As with comedy so with tragedy the titles of lost plays acted at court and preserved in Revels Office documents eke out our scanty knowledge of drama in the decade or so before Marlowe. But many of the titles are tantalizingly ambiguous. Was the tragedy of *The King of Scots* (Chapel children, 1568) historical or was it (as is more probable) as little historical as Greene's *Scottish History of James the Fourth* and based, perhaps, on Juan de Flores's romance concerning the ill-fated love between Aurelio and Isabel, daughter of the king of Scotland, translated into English in 1556? The two 'houses' or mansions supplied by the Revels Office—a great castle and the Palace of Prosperity—suggest a play that was in part morality. What was *The Cruelty of a Stepmother* about (Sussex's men, 1578), except the cruelty of a stepmother? And could *Murderous Michael* (Sussex's men, 1579) have been an early example of realistic tragedy antedating *Arden of Feversham* by more than a decade? In comparison *Iphigenia* (Paul's children, 1571) is plain sailing.

Among the lost plays of the fifteen-seventies are a surprising number based on classical history and legend: *Ajax and Ulysses* (Windsor boys, 1571); *Quintus Fabius* (Windsor, 1574); *Xerxes* (Windsor, 1575); *Mutius Scaevola* (Windsor and Chapel, 1577); *Alucius* (Chapel, 1577). If we bear in mind that the Revels records are by no means complete and fail us for 1572–3, 1575–6, 1577–8, and for 1574–5 except for the casual reference to 'king xerxces syster in ffarrantes playe', that such entries as 'the history of ' (Chapel, 1578–9) are useless, and that we have no information at all about plays which were not produced at court, this list is the more remarkable. All these plays were presented by one man. Some account of the children's companies will be given later,[1] but it must be said here that Richard Farrant became schoolmaster to the children of St. George's Chapel, Windsor, in 1564 and Master or deputy of the Children of the Chapel Royal, from 1576 till his death in 1581, and that in the last days of 1576 he acquired a lease of the first Blackfriars theatre. At this private theatre he began to

[1] Below, pp. 151 ff.

produce his children (whether of Windsor or the Chapel or both) for profit while continuing to present such performances as he was required to give at court. Farrant may himself have been the author of some of these plays: he was certainly trainer and producer of his boys, and obviously he had a taste for themes from classical history. It has been suggested that the theme of *Ajax and Ulysses* was the conflict between Ajax and Ulysses over the award of the armour of Achilles (Ovid, *Met.* xiii), that *Mutius Scaevola* was based on Livy (ii. 12–13) or Painter's translation, that *Quintus Fabius* was also based on Livy (viii. 30–35)—the theme turning on the necessity of absolute obedience to lawful authority, that Xerxes's sister was his sister-in-law, and the play about his violent passion for her daughter Artaÿnta (Herodotus, ix. 108–13), and that *A History of Alucius* concerned the betrothed of Alucius, prince of the Celtiberians (Livy, xxvi. 50), who was captured by Scipio Africanus and as honourably treated by him as was Lyly's Campaspe by Alexander.

Probably dramatist, Farrant was certainly choirmaster and composer, being today best remembered for his sacred music. A song of his, preserved at Christ Church, Oxford, is a lament sung by Panthea for the death of her husband 'Abradab'. The only surviving play based on the story of Panthea and Abradates, a story told in Xenophon's *Cyropaedia* (translated by W. Barkar, *c.* 1560), is *The Wars of Cyrus, King of Persia, against Antiochus, King of Assyria, with the tragical end of Panthaea.* It was printed in 1594, as played by the Children of the Chapel, in a poor text in which act divisions are confused, the prologue is misplaced, choruses are removed. The songs too have been left out, but enough indications survive to show that originally there were musical intervals between the acts, a characteristic of many later children's plays performed at private theatres.

The importance of the play hinges upon the date. Assign the composition to 1587–94 and the play falls in with *Alphonsus of Aragon* and *Selimus,* unimportant testimonies to the popularity of *Tamburlaine*. Assign it to *c.* 1577–80 and it becomes a significant pointer to the kind of play presented by Farrant at the Blackfriars in a theatre that had now become professional. If of early date, the play provides the only dramatic blank verse between the academic *Gorboduc* and *Jocasta* and the few specimens in Peele's *Arraignment of Paris*. It differs from *Horestes* and its like in that it has wholly abandoned allegorical characters and the allegorical mode. It has also abandoned the admixture of comic scenes: that this was deliberate is shown by the Prologue which as proudly announces a break from current dramatic fashion as does *Tamburlaine*:

> It is writ in sad and tragicke tearmes,
> May move you teares; then you content our muse,
> That scornes to trouble you againe with toies
> Or needlesse antickes, imitations,
> Or shewes, or new devises sprung a late.
> We have exilde them from our Tragicke stage,
> As trash of their tradition that can bring
> Nor instance nor excuse for what they do.
> Instead, of mournefull plaints our *Chorus* sings:
> Although it be against the upstart guise,
> Yet, warranted by grave antiquitie,
> We will revive the which hath long beene done.

That Farrant wrote a dirge sung by Panthaea for the death of her husband is powerful evidence that the extant play was written before his death in 1581. Supporting evidence is that the Children of the Chapel, to whom the play is ascribed on the title of 1594, were not acting in London between 1584 and 1600. Again, there seems to be some evidence that 'shews' were not so popular after 1578 as they were before. At Trinity, Cambridge, in the fifties and sixties many shows were given: after 1570 they are rarely mentioned. And in the Revels Accounts while down to 1578 the Office was furnishing 'plaies comodies or shewes of histories and other Inventions and devises incident', after that date the word 'shews' is dropped, except for a reference to 'feates of activity and other shewes' in 1587. We cannot suppose that dumb-shows are meant, only incidental ornaments to tragedy. The distinction seems to be between the traditional forms of comedy and tragedy and an entertainment that resembled a medieval disguising, one in which spectacle, dancing, and miming were more important than dialogue.

But all this evidence goes for nothing if it is maintained that *The Wars of Cyrus* is an imitation of *Tamburlaine*. In comparison the style is quiet and pedestrian: it is without the bombast, as it is without the genius. If an occasional piece of description or hyperbole reminds us of Marlowe the reason may be that both dramatists in their different ways are trying to create an atmosphere of oriental splendour. The short-title *The Wars of Cyrus* suggests that the play is another conqueror play, and it does make the contrast between the good conqueror Cyrus and the evil Antiochus and show the triumph of Cyrus in battle. But the heart of the play lies in the last words of the title—'With the tragical ende of Panthaea'. Panthea's husband appears only momentarily in the fifth act; Panthea is present in every act, and is sometimes present even when absent. We see her as a captive honourably entertained by Cyrus (as was the affianced of Alutius by Scipio Africanus, as Campaspe by Alexander); we see her solicited by Araspas, the soldier appointed by Cyrus to be her guardian, and we watch her repudiation of him and Cyrus's anger at Araspas; we see her welcoming and arming her husband and after his death in battle witness her grief as she stabs herself 'for Abradates sake' by the 'sad and hollow bankes' of the Euphrates. The end is Cyrus's tribute to her virtues and her husband's. Here is pure tragedy of fortune: these are star-crossed lovers.

Much of this is in Xenophon, but not all. In his 'Third Blast' of 1580 Anthony Munday attacked the dramatists who gave known histories 'a new face' so making that which was old seem new. This dramatist gave his history a new face by heightening the romantic interest. Alexandra, captive of the cruel Antiochus, whose life is saved by an exchange of clothes with her heroic page, is wholly his invention. Also invented are the scenes in which Araspas attempts to win Panthea's love by magic—vainly, for magic cannot command the soul and must yield to virtue. Adding to the seriousness of the play and again not in the source are the occasional halts in the action when some moral or philosophical issue is briefly debated, as when Panthea maintains that the perturbations that assail the mind can be mastered even if chance may not be—'My minde and honour free and ever shall'; or as when Cyrus denies that love is voluntary and refuses at first to see Panthea:

> Nothing can more dishonour warriours
> Then to be conquered with a womans looke.

*The Wars of Cyrus* is a poor thing to set beside Marlowe, yet when we consider that much work of its kind is lost to us, work in which serious themes were seriously treated, we must concede that theatres which could offer such wares might tempt a promising young man to turn to them. The theatres were ready, so was the audience. A form for comedy and a form for tragedy had been evolved. It was no longer necessary for a dramatist to write in fourteeners or to follow the 'jigging veins of rhyming mother wits'. But what no one had as yet done, except occasionally in song, was to bring to the theatre 'poetical sinews'. These tragedy demands even more than comedy, and now at last, when the reign of Elizabeth was half over, the theatre was to be supplied with a succession of young men, some from the universities and some not, who were not only dramatists but dramatic poets.

# VII

## THE THEATRE BEFORE 1585

THE purpose of this chapter is to trace the history of the chief dramatic companies up to 1585, to consider the places in which they acted until, and for a short time after, they built or secured their own theatres, and to examine such evidence as remains of the Revels Office and its many activities. The period is that in which the professional companies of adult players became increasingly important until in the fifteen-seventies they outstripped the boy companies. The section will conclude with a brief account of the hostility to common players shown by many a preacher from about 1577, the stringent regulations which the City of London attempted to enforce, and the interventions of the Privy Council which saved the players from total suppression.

The story begins of right with the boy companies, for they were the first who secured the favour of the court and, outside the universities and the Inns of Court, best maintained in this period the standards of the drama. The development of these boys into companies of professional actors is not to be paralleled before or since. It becomes more intelligible if we remember that from the fifteenth century boys had taken their part in royal and civic pageantry, that in the ceremony of the Boy Bishop every Innocents' Day, a ceremony which persisted till England became Protestant, the public had become accustomed to seeing children act a role which was in part mimetic, that most of them were choir boys specially picked and carefully trained, and that the masters who picked and trained them were sometimes men of unusual ability and energy. We may also remember that in the early years of these companies there was little competition from the despised common players and there was no *Hamlet* or *King Lear* to show up their limitations.

While the history of the Chapel as a part of the Royal Household goes back to the twelfth century, the Children of the Chapel are at first mentioned for their musical accomplishments:

they were choristers before they were actors, as we may see from the elaborate disguisings at the wedding of Prince Arthur and Catherine of Aragon in 1501. Under the vigorous mastership of William Cornish[1] (1509–23), however, the children took to the stage. Much depended on the master who had charge of the children and from early days possessed an authority to recruit almost as drastic as that of a press-gang. Cornish, musician and composer and entrepreneur, was especially enterprising and gifted as organizer of the sumptuous disguisings in which the court of the young Henry VIII delighted. No part of his dramatic writings is extant, but that he was responsible for turning his boys from song to drama is certain, and the titles of a few of the plays which he produced and probably wrote have survived. Succeeding masters also took their boys to play before the Queen, receiving for each occasion the standard fee of £6. 13*s*. 4*d*. (ten marks). The fee remained constant, for children and men alike, till 1575–6 when it was raised to £10 if the Queen was present.

Passing over the puzzling share taken by John Heywood and Nicholas Udall in the production of plays at court,[2] neither of them masters of a children's company, we come to the distinguished mastership of Richard Edwards (1561–5).[3] Here at last is a master of whose work we have first-hand knowledge. In 1565 and 1566 he took his boys to act at Lincoln's Inn, the earliest instance of the Chapel boys acting elsewhere than at court. He was succeeded by William Hunnis (1566–97) known to us as the author of *Seven sobs of a sorrowful soul for sin* (1583) and other pious verses; but from 1576 till his death in 1580 the effective master in the production of plays was Richard Farrant, Master of the Children of the Chapel Royal at Windsor, but also a Gentleman of the Chapel. For the Windsor boys he had produced a play at court once a year from 1566–7 till 1575–6; then in the last four years of his life the Children of the Chapel, strengthened perhaps by the Windsor boys, acted under his direction an interesting series of historical plays. Farrant's initiative in securing for the Chapel a theatre of its own adds more distinction to his short rule.[4] After his death Hunnis took charge once more. Large claims have been made for him as a dramatist, as also for William Cornish, but in the absence of all

[1] Above, pp. 22 f. [2] Above, pp. 32, 104. [3] Above, pp. 111 f.
[4] Above, pp. 146 ff.

evidence their plays must excite an interest which cannot be satisfied. Presumably it was Hunnis who produced his boys in Peele's *Arraignment of Paris* in 1584 or earlier. Between 1584 and 1600 the Chapel children are not known to have acted publicly in London or at court.

Less is known of the Chapel children's rivals, the children of Paul's, that is, of the cathedral's song school not of Colet's day school. Of John Redford, master from 1531–4 till his death in 1547, something has already been said.[1] No doubt his boys acted his *Wit and Science* and assuredly with applause. He was succeeded by his friend and literary executor Sebastian Westcott, the Papist whom Grindal did not succeed in ousting from the mastership.[2] In 1551–2 we find him receiving payment for a performance before Princess Elizabeth, and from 1560 (when the records begin of payments for performances at court in the annual declared accounts of the Treasurer of the Chamber) down to the year of his death 1582, only one year (1569–70) passes without Westcott receiving payment for presenting plays at court: in seven seasons he presented two plays a year and in 1565 three. It is a remarkable record, not to be matched by the Chapel children or any men's company during these years. There can be no doubt that he and his boys had the ear of the court, and it is easy to see why Elizabeth refused to dismiss him from his post, though a Papist. And yet, although Westcott's boys gave twenty-eight performances during these years, we cannot with certainty identify a single play that has survived, a striking instance of how slight our knowledge is of these formative years. After 1582 the Paul's children disappear from court for a few years, and a mysterious company called Oxford's children begin to act in 1584. Brief as was its career and little as we know of it, it is of great importance, for these were John Lyly's boys.

The dramatic activities of some of the grammar schools have already been mentioned: Magdalen College School about 1509–10 and Sir Thomas More's interest in it; Colet's St. Paul's under John Rightwise which probably presented the Latin morality acted in 1527 before Henry VIII and a distinguished audience; Hitchin School under Bale's friend Ralph Radclyff who built a stage in his school, *c.* 1538 and taught his boys to speak Latin clearly and elegantly in a variety of plays, some biblical, some (*Patient Griselis, Titus and Gisippus*) not, and some

[1] Above, p 43. [2] Above, p. 105.

(Chaucer's *Meliboeus*) moral; Bishop Gardiner's School at Southwark for which Udall may have written *Roister Doister*.[1] An open-air amphitheatre near the Severn was used at Whitsuntide by Thomas Ashton, the distinguished Headmaster of Shrewsbury School from 1561, for the performance of plays on religious themes (*Julian the Apostate*, *Passion of Christ*). The plays seem to have belonged more to the town than to the school, but many of the actors Ashton doubtless chose from his school. In the performance of Richard Legge's *Richardus Tertius* at Ashton's old college of St. John's in 1579 no less than five boys from Shrewsbury, among them Abraham Fraunce, appear as actors. It was in 1564 that Ashton admitted to the school Philip Sidney and his cousin Fulke Greville.

After the rise of the Chapel and Paul's children these grammar school performances became less important. That they continue is shown by words which appear in a Fancy of street cries set to music by Richard Deering early in the next century:

O yes, all that can sing and say,
Come to the town hall, and there shalbe a play,
Made by the Schollers of the Free Schoole,
At six a clocke it shall begin,
And you bring not money you come not in,

but these do not affect the current of national drama. We may note, however, that the boys of Westminster acted Terence and Plautus, *Sapientia Salomonis*, *Paris and Vienne* at various dates between 1564 and 1572, that the boys of Eton were at court in 1573, and that the boys of Merchant Taylors, trained and produced by Richard Mulcaster, acted before the Queen five times between 1572 and 1576 and once more in 1583 in plays of which a few titles only are left to us. In this way, testifies one of his old actor-pupils, he taught his boys 'good behaviour and audacitye'. 'Audacitye' was a good commodity to have when playing before the Queen.

During the mastership of Sebastian Westcott and the headmastership of Richard Mulcaster the men's companies for the first time outstrip the boys' companies in favour at court. From about 1575–6 they play more often before the Queen: in 1577–8 the proportion is as high as nine (mainly Leicester's, Lord Chamberlain Sussex's, and Warwick's men) to two (Chapel and Paul's). The increase is in part due to the proliferation of

[1] Above, p. 108.

men's companies especially about the time (1576–7) of the building of the Theatre and the Curtain.

The last quarter of the century witnessed the triumph of the professional actor. In the first half of the century there is often difficulty in distinguishing between the professional and the amateur or semi-amateur:[1] in the last quarter of the century the distinction has become clearly marked. Yet professionals there had been in the later Middle Ages. Attention has already been drawn to the players in the household of the Earl of Northumberland.[2] Henry VII and his successors also supported players; these 'players of the king's interludes' remained effective in the royal household till Mary's reign. Under Elizabeth they became mere pensioners, and the last of them died off in 1580. Their names can be recovered from the Public Record Office, but who remembers them? How many can remember the names of any professional English actors before the fifteen-seventies? In that decade emerged Tarlton and Wilson, and soon afterwards Edward Alleyn our English Roscius and Richard Burbage, and since that time the supply of famous actors has never failed.

The earliest legislation concerning plays was confined to the censorship of their contents, a statute of 1543 prohibiting any plays which conflicted with authorized religion. There were similar enactments under Edward VI and Mary. A system of local censorship was contrived which cannot have been effective, and not until after 1580, when the drama had become more centralized in and about London and the Master of the Revels had been given greater powers, did censorship begin to work tolerably well. In the licensing of actors the government was more successful. Players under the protection of some noble person were in no danger of being treated as rogues and vagabonds, masterless men to be whipped from parish to parish. The patrons themselves were sometimes held responsible for the conduct of their players. The various Elizabethan poor laws culminated in the acts of 1572 (14 Eliz., c. 5) and 1597–8 (39 Eliz., c. 4) by which only barons or persons of greater degree might license players who 'wandered abroad'. The law did not apply to those who supported a company of players who did not travel or travelled only to play in a neighbour's house, and in the sixteenth century we hear of country gentlemen like

[1] Above, pp. 50–52. [2] Above, pp. 1 f.

Alexander Houghton of Lea, Lancashire (d. 1581), who kept players and play clothes. Among his servants was a William Shakeshafte whom some would persuade us to think of as William Shakespeare errant in his boyhood from Stratford-upon-Avon.

An early example of a players' licence was shown in 1583 to the town officials of Leicester by the Earl of Worcester's company. The Earl named the servants whom he had licensed 'to play and goe abrode, usinge themselves orderly' and required that they should be given such entertainment as other noblemen's players had. In 1574 the Earl of Leicester's company received the more signal honour of a licence by royal patent, the only Elizabethan example. It demands in the Queen's name freedom to act in London and elsewhere without hindrance or molestation, any former act, statute, or proclamation notwithstanding, provided that their plays had been allowed by the Master of the Revels and were not acted in time of common prayer or in plague time in London. Among the six actors named are Robert Wilson and James Burbage. Piety demands a pause at the first mention of a Burbage, for the debt of the Elizabethan drama and theatre to James and to his sons Richard and Cuthbert is inestimable. Many years later Cuthbert claimed that the success of the Globe and the Blackfriars was 'purchased by the infinite cost and paynes of the family of the Burbages, and the great desert of Richard Burbage for his quality of playing'.

No doubt patrons varied in the amount of attention they paid to the companies under their protection. Some were fickle and indifferent, others (and Leicester among them) seem to have taken some trouble in promoting their interests. We find him as early as 1559 recommending the Earl of Shrewsbury, President of the North, to allow his players to act in Yorkshire and testifying to their honesty and to the 'tollerable and convenient' nature of their plays; and the exceptional honour of a royal patent must be ascribed to him. Nor did his influence count for nothing in securing for them a hearing at court. Their great years were from 1572–3 to 1582–3 when they acted eighteen times. When the members of the company, alarmed by a proclamation of 1572 limiting the number of retainers, petitioned him to make them his household servants (without stipend) and not merely his liveried retainers, they rightly addressed him as 'their good lord and master' and ended in a burst of verse. We may surmise that without confidence in the protection of his powerful patron James Burbage might not have taken the daring step of building a permanent theatre in 1576.

The Earl of Sussex's men were also fortunate in their patron. Lord Chamberlain from 1572 till his death in 1583, he was responsible for the choice of court plays, and during those years his players made fourteen appearances. But of all the early Elizabethan companies the most important was that formed under the patronage of the Queen herself. Edmund Howes the chronicler writes in 1615:

> Comedians and stage-players of former time were very poor and ignorant in respect of these of this time: but being now grown very skilful and exquisite for all matters, they were entertained into the service of divers great lords: out of which companies there were twelve of the best chosen, and, at the request of Sir Francis Walsingham, they were sworn the queens servants and were allowed wages and liveries as grooms of the chamber: and until this yeare 1583, the queene had no players. Among these twelve players were two rare men, viz. Thomas [*sic for* Robert] Wilson, for a quicke, delicate, refined, extemporall witt, and Richard Tarleton, for a wondrous plentifull pleasant extemporall wit, he was the wonder of his time. [*margin*] He was so beloved that men use his picture for their signs.

Wilson we have already met, and also his fellow Richard Tarlton (d. 1588), the greatest of Elizabethan clowns.[1] Between 1583–4 and 1590 the Queen's men never failed to act at court twice a year and sometimes four or five times a year. Wilson and Tarlton and their fellows in the 'quality' were no vagrants but grooms of the Queen's Chamber, wearing the royal livery; and from 1583 till the closing of the theatres London was never without at least one company of professional actors with a stake in the country and decent standards of craftsmanship. Touring the country towns and large country houses in the summer, rehearsing and performing their new plays in the autumn in the London suburbs, in the winter moving nearer to the heart of London (when the City could not keep them out) and with that shining goal before them—the glory and the profit of acting before the court itself—they were more in touch with the nation at all levels of taste and

[1] Above, p. 58.

intelligence, and with all classes of society, in city, court, and country, than any English actors at any other time.

Actors are an adaptable race, and in Elizabethan times they needed to be. At one extreme we have the elaborate performance of an accredited company at court with the wealth of the Office of the Revels and its stock of scenery and costumes behind them, and at the other a performance in the country by a strolling company stalking 'upon boords, and barrell heads, to an old crackt trumpet'. Actors might have to perform in the hall of an Inn of Court, in the private house of a lord or gentleman or city magnate, in a garden house, in inn-yards, and in the country on many a makeshift stage in schoolhouses, parish halls, brewhouses, and even in churches. But though so adaptable every actor sighs for a fixed abode. The only Elizabethan building which could and did call itself The Theatre was that built in Shoreditch in 1576 by James Burbage, the first public theatre in England specially built for the production of plays. Years earlier two rings had been built on the Southwark side of the Thames for bull and bear baiting, with roofed seats for the spectators but no roofs for the bulls and bears and their baiters, so that Burbage's building came late rather than soon. But it came.

From 1557 we hear of the inn-yards of London as congenial playing places, and some like the Bull in Bishopsgate Street, and the Bel Savage on Ludgate Hill have already been mentioned.[1] In 1576[2] William Lambarde in his *Perambulation of Kent* asserted that pilgrims to Boxley Abbey and the miracles of St. Rumwald could not assure themselves of any good gains without a treble oblation:

> No more then suche as goe to Parisgardein, the Bell Savage, or some other suche common place, to beholde Beare bayting, Enterludes, or Fence playe, can account of any pleasant spectacle, unlesse they first paye one penny at the gate, another at the entrie of the Scaffolde, and the thirde for a quiet standing.

An inn with an archway at which money could be collected leading to a galleried courtyard provided a stage, a 'yard' for the audience, an upper level for the actors (when the play demanded it) and select spectators, and a tiring house. But there

[1] Above, p. 118.

[2] E. K. Chambers (ii. 359) says that the passage is not in the first edition of 1576; but it is (at p. 187), though instead of 'Sauage . . . beholde' 1596 reads '*Sauage,* or *Theatre,* to beholde'.

were obvious disadvantages. Stages had to be temporary in all but the few inns which specialized in plays, landlords might prove extortionate, accommodation was limited and sometimes money must have been turned away, and the inns lay mostly in parishes that came under the control of the watchful and increasingly hostile City of London. But the justices of the peace of Middlesex and Surrey under whose jurisdiction came the outparishes and some of the liberties were either negligent or tolerant, and Burbage on borrowed capital and with a lease of twenty-one years wisely put up his building in the liberty of Holywell, part of the Middlesex parish of St. Leonard's, Shoreditch, close to Finsbury Fields. With Burbage's sharp financial practices and the lawsuits to which they led, with young Richard's disdainful playing with a deponent's nose, we are not concerned. In 1576 James was still, it would seem, one of Leicester's players, but later his theatre was used by other companies, by the Queen's men among others. The Curtain also was not the monopoly of any one company, and was outside the City's control. Built very soon after the Theatre, it lay in Moorfields a little south of the Theatre. The only other theatre built before 1585 was at Newington Butts, a mile from London Bridge. Perhaps it was too remote to catch on: but all the later public theatres of fame, except the Fortune and the Red Bull, were built south of the river within the lenient rule of the Surrey justices.

Late in 1576 and inspired perhaps by Burbage's example, Richard Farrant established the first of the so-called private theatres by converting the frater in the old Priory buildings of the Black Friars into 'a continual house for plays'. Although within the city walls, which formed a north and north-west boundary to the priory, the Blackfriars precinct remained after the suppression a liberty which disputed the jurisdiction of the City, nor was the dispute settled in favour of the City until 1608, and then only for petty offences and the keeping of the peace. 'Private' theatres, like the 'public' schools of today, charged for admission and catered for a privileged class. Unlike the Theatre and the Curtain, Farrant's converted frater was roofed (and therefore weather-proof) and lit by candlelight and torches; also its accommodation was limited, it provided seats for all, and it charged higher prices. His theatre and the second Blackfriars of 1596 were the only sixteenth-century private theatres, and not until 1608 when the King's players acquired the use of the second Blackfriars did a men's company act in a private theatre. The Paul's children, it is supposed, continued to act in their own singing school, and like the Chapel children played before a paying public.

Of the interior structure of the Theatre and the Curtain and especially of the stage and its exits and entrances we know little. Some will say that we are not much better off in respect of the Globe and the Fortune. The interiors of the two earliest theatres had an unroofed 'yard' (admission one penny) where the groundlings stood, seats were provided in the three rising galleries for spectators willing to pay more, the stage was a platform jutting out into the auditorium and surrounded on at least three sides by spectators. If as seems likely Shakespeare's *Henry V* was one of the last plays acted at the Curtain before his company moved to the Globe, then that theatre had a circular auditorium ('this wooden O'). *Romeo and Juliet*, said by Marston in 1598 to have received 'Curtaine plaudeties', was acted on a stage which somehow supplied an upper level for Juliet's bedroom and for the tomb of the Capulets a space concealed or revealed by a curtain. While the inn-yard provided a model for the general structure of the new theatres, recent writers have argued that we should look elsewhere for a model for Shakespeare's stage. For two centuries the people had been familiar with stages which could hardly have failed to influence Burbage—the stationary pageants built for royal and civil progresses and the pageants sometimes processional and sometimes stationary of the miracle plays, both with two or three levels. That plays could indeed be acted on street pageants or some of them is shown by the performance of 'a goodly Stage Play' on one of those which greeted Henry VIII at Coventry in 1511. Burbage did not have behind him the financial resources of court and city, but no doubt he expended much paint to make his theatre as gay as a pageant, the use of curtains and the mechanism of the windlass by which beings could be made to ascend and descend were known to him as to the Middle Ages, and clouds, blazing stars, apparitions of suns and moons came well within his compass. But the extent of our knowledge and of our ignorance of staging in public theatres cannot be explored here.

Of staging at court, especially of the costumes and properties used, we know much, thanks to the detailed declared accounts of the Revels Office which survive for some years of Elizabeth's reign. The first permanent Master of the Revels was Sir Thomas Cawarden (1545–59), the most distinguished of the sixteenth-century Masters Edmund Tilney (1578–1610). The Master was chiefly responsible to the Lord Chamberlain for the court's entertainment, not merely in respect of plays but of 'shews', feats of activity, challenges at tilt, and fighting at barriers. He led an active life, for he had frequently to wait upon the Lord Chamberlain, wherever the court lay, to submit plans for masques and the manuscripts of plays. In later years Tilney became censor of all plays, whether for the stage or the press, and if for the stage then whether for performance at court or not, but at all times the master must have acted under the Lord Chamberlain as censor of all plays to be acted at court. To him fell the initial choice and the alteration in the play chosen of anything that might be offensive or (a grave fault) tedious. The 'connynge' of the Office, wrote an unidentified officer in 1573,

> resteth in skill of devise, in understandinge of historyes, in judgement of comedies, tragedyes and shewes, in sight of perspective and architecture, some smacke of geometrye and other thinges wherefore the best helpe for the officers is to make good choyce of cunynge artificers severally accordinge to their best qualitie, and for one man to allowe of an other mans invencion as it is worthie, *especiallye to understande the Princes vayne.*

The last words deserve to be italicized. The influence of a Queen, and such a Queen, upon the drama has not been treated only because it cannot be estimated; but we may be sure she was an exacting critic not slow to express pleasure or displeasure. The chief duty of the master was to give pleasure, and to do that he had to take pains and precautions. So we find in the commission of 1578 defining Tilney's powers as Master that he might call all players and their playmakers, whether belonging to a nobleman or not, to recite before him or his deputy any tragedies, comedies, interludes, or other shows they had in readiness in order that they might be authorized, reformed, or put down. So in 1571–2 six plays were chosen out of many 'and ffownde to be the best that then were to be had, the same also being often perused, and necessarily corrected and amended'. Rehearsals appear to have been held at night, after the players' usual playing time, at the Revels Office in St. John's Gate, Clerkenwell, and possibly also at court. Children as well as men were carefully rehearsed. The Terence and Plautus acted before the Queen by the grammar boys of Westminster in 1564 were rehearsed before the Master of the Revels; the nine children who took part in the Shrove Tuesday masque of 1574 were taught their 'partes and Iestures'; and in preparation for Christmas 1578 the Chapel children were taken by boat to Richmond 'to Recite before my Lord Chamberleyne'. If the Office's accounts represent the facts, no pains were spared in the attempt to provide finished performances and to please the Queen's 'vein'. The Master of the Revels to Shakespeare's Theseus showed the same scrutinizing care, the rehearsal of Bottom's play reducing him to tears—of laughter.

Not the least arduous of the duties of the Office was the maintenance of a wardrobe. This duty became less arduous during Tilney's mastership when for reasons of economy the needs of the Office were supplied more and more by the Great Wardrobe, but for many years in Elizabeth's reign the Revels give full details of the expenditure on garments for plays and masques. Even professional players acting at court in these early years were supplied with clothes, as if their own were not resplendent enough to appear before the Queen. John Holt, yeoman of the Office (d. 1571), went so far as to say that the accounts of the Office 'hath bene but a Taylers Bill', but he exaggerated. Tilney's patent authorized him to take and retain at competent wages as many painters, embroiderers, tailors, cappers, haberdashers, joiners, carders, glaziers, armourers, basketmakers, skinners, sadlers, waggon-makers, plasterers, feltmakers, and other property-makers and cunning artificers as he should think necessary. Among the other cunning artificers were mercers, wiredrawers, silkmen, carpenters, plumbers, cutlers, feathermakers, drapers, furriers, and the John Izard who invented a device for counterfeiting thunder and lightning in the play of *Narcissus* (1572). (He was paid 22*s.*, whereas Westminster School playing the *Mostellaria* in 1569 paid only 2*s.* for the loan of a thunder barrel and for the man who 'thondered' but may not have provided lightning.) The detail shown in some of these declared accounts is such that in 1574–5 the clerk of the Revels momentarily lost patience and ended a long itemization of the money owing to a deceased property-maker with the words 'Nayles, hoopes, horstailes, dishes for devells eyes, heaven, hell and the devell and all, the devill I should saie but not all'.

These accounts, so full for many years under Elizabeth, fail us after 1589 just at the time when we should most wish to consult them. In matters of censorship the Office became more and more important as more power was granted to the master, but otherwise its importance dwindled. The Wardrobe, as we have seen, took over some of its duties, and so did the Office of Works, an Office responsible for the upkeep of the Queen's standing palaces, such as Whitehall, Greenwich, Hampton Court, and Richmond. England had no court theatre until Inigo Jones built his Cockpit-in-Court at Whitehall in 1630, and an *ad hoc* stage had to be built every time a masque or play was performed at court. For the building and dismantling of these temporary stages or platforms and of the galleries and seats for spectators the Office of Works was responsible both in this period and later, and its declared accounts usually tell us in which hall or room of which palace plays were acted, and in Jacobean and Caroline times tell us something of the scenery and machinery required for the court masques. Before 1589, however, some of the work later done by the Office of Works was done by the Revels Office. In 1571–2 not only were properties supplied for the six plays chosen but also 'apt howses: made of Canvasse, fframed, ffashioned and painted accordingly', and in other years the Revels supplied cities, chariots, rocks, fountains, arbours, castles, battlements, pulleys for clouds and curtains, monsters, mountains, forests, and dragons.

The expenditure on plays as distinguished from masques was perhaps never greater than during these years. The considerable resources of the Revels Office were lavished upon the productions of children's and men's companies; and there was a noticeable increase in the number of plays acted at court during the great festal days of the year—the twelve days of Christmas with special emphasis on St. Stephen's Day (26 December), St. John the Apostle's (27 December), Holy Innocents' (28 December), New Year's Day, and Twelfth Night (6 January), Candlemas (2 February), and Shrovetide. Add the other developments favourable to the actors—the patronage of the nobility and, in the case of one company, of the Queen herself, the establishment of one private and two public theatres—and we might conclude that the players were on the top of golden hours. Yet at this time their very existence was threatened.

Perhaps the stage has never been without its enemies in England: a fourteenth-century priest attacked all miracle plays for making 'play and bourde' of the miracles and works of God.[1] But before the fifteen-seventies attacks were few and far between and from that date until the closing of the theatres in 1642 the common stages were continually being assailed. Before the seventies professional drama was presumably of so little account that the moralists could afford to ignore it: but by the late seventies, with three permanent theatres opened in London and several other playing places, the increasing popularity, secularization, and commercialization of the drama could no longer be ignored by magistrate or preacher. Academic drama, too, begins to be found reprehensible by some. So long as the humanistic faith in the moral and educational uses of play-acting persisted, it was let alone, but by the seventies academic dramatists had cast their net so wide as to include subjects which were more entertaining than edifying.

An early attack on common players appears in *A Form of Christian Policy* 'gathered out of French' by Geoffrey Fenton in 1574. It anticipates many of the charges made by later writers: the infamy of players in ancient Rome; the hostility of the primitive Church; the corruption of good instruction (when any is offered) with 'Iestures of scurrilitie', interluders babbling vain news on a scaffold and scoffing at the virtues of honest men, so that it is impossible to draw any profit out of the doctrines of their spiritual moralities; the impiety of acting on the Sabbath; the majesty of God often offended in the two or three hours the plays last. Academic drama is excepted, provided that it is not based on 'the superstitions of the Gentiles', rebukes vice, and praises virtue, and encourages young scholars in boldness of speech in all honourable companies and in a well-disposed eloquence. More effective were the attacks made by Anglican ministers of a Calvinistic persuasion preaching at Paul's Cross, the great open-air pulpit of the cathedral and indeed of the Church of England. On 3 November 1577, in a sermon preached in time of plague, Thomas White drew attention to 'the sumptuous Theatre houses, a continuall monument of Londons

[1] Above, p. 60.

prodigalitie and folly', and proved by logic that plays were the direct cause of plagues; for 'the cause of plagues is sinne, . . . and the cause of sinne are playes: therefore the cause of plagues are playes'. A month later a pamphlet was entered in the Stationers' Register which attacked 'Vaine playes, or Enterluds, with other idle pastimes, &c., commonly used on the Sabboth day'. In this scissors-and-paste compilation John Northbrooke, a Gloucester minister, names and blames the Theatre and the Curtain for the filth which the players utter. In the next year and again in 1579 John Stockwood, schoolmaster of Tonbridge, referred at Paul's Cross to that 'gorgeous Playing place' the Theatre and seven other playing places, all so popular that if they acted only once a week they took in about £2,000 a year. Like many another he was indignant that performances should be given on Sundays in time of divine service so that many churches were empty when the theatres were 'as ful as they can throng'. The most effective onslaught, however, came from a playwright turned pamphleteer whose works have already been discussed, Stephen Gosson in his *School of Abuse* and *Apology of the School of Abuse* (1579).[1] An able writer and controversialist, Gosson attacked the immorality of the plays acted at the new playhouses and the immorality of the audiences who attended them, 'a generall Market of Bawdrie'.

Gosson was answered on behalf of the players by Thomas Lodge, an Oxford graduate and student of Lincoln's Inn, but *Honest Excuses* is a feeble compilation written by a young man pretending to more learning than he had, and Gosson had no trouble at all in destroying his arguments in his *Plays Confuted* (1582), the interest of which to us today lies in what it tells us of that dark period of the drama from about 1576 when so much was happening that we should like to know about. Following on Gosson's 'first blast' came *A second and third blast of retrait from plays and theatres* (1580). The second blast is a translation from a medieval bishop, the third is Anthony Munday's. Munday, like Gosson, had been a playwright (so he says), and unlike Gosson was to return to the theatre. He pleaded for wholesale extirpation, but short of that suggested that Sunday performances should be prohibited, that the patronage extended to players by the nobility should be restricted, and that the practice of training young boys to act should be prohibited.

[1] Above, p. 118.

His pamphlet carries the arms of the City of London, and the City, it has been suggested, may have commissioned him, for he was a man willing to turn his pen to almost any purpose: certainly, his doctrine would meet with their approbation. One more attack out of many may be mentioned by reason of the book's fame. In a preface to the first edition of *The Anatomy of Abuses* (the imprint dated 1 May 1583) Philip Stubbes allowed that some kinds of plays were 'very honest and very commendable exercyses' and acknowledged that what we see makes a deeper impression than what we hear; but the preface is omitted from the later editions and there is nothing to mitigate his charge that the public theatres are palaces of Venus and schools of mischief, and that those who frequented them were incurring the danger of eternal damnation.

The attention of these early attackers was concentrated on the public theatres. Fenton (or his original) and Northbrooke are almost alone in mentioning academic drama. It was lawful, Northbrooke granted, for a schoolmaster to train his scholars to play comedies, but he insisted on these provisos: no wanton toys of love and no ribaldry and filthy terms; acting for the sake of learning and 'utterance', usually in Latin and very seldom in English; performances to be 'verye rare and seldome'; no gorgeous apparel; and no acting publicly for profit. Some of the objections to common players could not reasonably be applied to university players, and the hostility to academic drama depended mainly upon the belief that all plays were evil, even (or, rather, especially) the visible representation of sacred themes. It is a little ironical that two undergraduates noted in later life for the severity of their principles did not scruple to act in university plays. There is Thomas Cartwright of Trinity, Cambridge, who in 1564 produced the *Trinummus* of Plautus and charmed Elizabeth with his acting in Haliwell's *Dido*. And there is John Rainolds of Corpus, Oxford, who lived to regret that in 1566 he played a woman's part (Hippolyta) in Edwards's *Palamon and Arcyte*, even though he did perform it before the Queen at Christ Church.

The controversy between Rainolds (who had been Gosson's tutor at Corpus) and the Christ Church dramatist William Gager is much the most interesting episode in the academic stage quarrel. Rainolds had been invited to attend the performance of three Latin plays by Gager in February 1592, sent a civil refusal, and later a statement of his objections. Into the third of these plays Gager introduced a character Momus who was ridiculed for his extreme hostility to plays in general. Unjustly but not unnaturally Rainolds believed the satire to be personal, and in two letters to Gager, dated 10 July 1592 and 30 May 1593 and both printed in *Th'Overthrow of Stage-Plays* (1599), he expounded his views at length. Some of his points had already been urged by his pupil Gosson and others. He objected to men wearing women's apparel, basing himself mainly on Deuteronomy xxii. 5. He objected to Sunday performances as a profanation of the Sabbath, a view still taken by the law in Great Britain; that one of the plays to which he had been invited was performed on a Sunday added fuel to his fire. The waste of time that might be more profitably employed, the expenditure of money that might be more usefully spent, the notorious infamy of actors, the lewd speeches and drunken gestures which they were forced to utter or feign, were other objections. Rainolds's arguments are pressed with sincerity, force, and learning, and granted his premisses most of them are unshakable. In a good-tempered letter dated 31 July 1592 Gager gave his answer. He denied that any just comparison could be drawn between professional actors whether of Rome or England and the young men who made their rare appearances in college plays and acted without gain, between those who corrupted the manners of their audience with prodigal humours and those who acted to recreate their college or university with some learned poem or other. In insisting that university actors were amateurs he recalled how a student of his college when he 'should have made a *Conge* like a woman, he made a legg like a man'. He maintained that they cannot be said to 'wear' women's dress who use it only for a few hours and with no intention to deceive, a quibble not so convincing as the explanation of John Selden, which converted William Crashaw, that the text in Deuteronomy applied to the ceremonial not to the moral law. He insisted upon the beneficial effects of good plays and the lessons to be drawn even from the representation of drunkenness upon the stage, an argument which would have appealed to Sidney, and to the charge of extravagance he replied that if his college spent thirty pounds on plays once in many years the poor might yet be relieved. As for the charge of profaning the Sabbath he pointed out that his play was acted on

Sunday night, not in time of divine service, and—a debating point—that those who misliked the play and did not attend may well have been worse occupied than were the actors. Memorable are the words in which he states the ends and effects of acting in universities:

to practyse owre owne style eyther in prose or verse; to be well acquaynted with *Seneca* or *Plautus*; . . . to trye their voyces and confirme their memoryes; to frame their speeche; to conforme them to convenient action; to trye what mettell is in evrye one, and of what disposition thay are of; wherby never any one amongst us, that I knowe was made the worse, many have byn muche the better.

Years later that eminent lawyer Francis Bacon (*De Augmentis*, vi. 4) held that stage-playing was an essential part of a young man's education because it strengthened the memory, regulated the tone and effect of the voice, taught a decent carriage of the countenance and gesture, begat no small degree of confidence, and accustomed young men to bear being looked at.

The university authorities and the supporters of academic drama were united at this time in their opposition to the common players: in the hey-day of Shakespeare and Jonson they were to become more tolerant. The common players, we may suppose, were not much disturbed by the attacks made upon their characters, their morals, their plays and their audiences provided that all this agitation did not deprive them of their livelihood. But this danger was serious. In the early years of the century many of the City of London's injunctions seem to refer to amateur performances, no doubt of miracle plays, saints' plays and moralities. Sometimes the City gave an exclusive permit to a parish to produce a play for so many months, the proceeds to be used for the repair of a parish church. In 1542 they prohibited all plays till further notice, probably in reference to the complaints that players were contemning God's word: according to W. Turner (*The Rescuing of the Romish Fox*, 1545) Bishop Gardiner forbade the players of London to play any more plays of Christ and confined them to trifles about Robin Hood and the Parliament of Birds. A proclamation of 1545 condemned Sunday acting especially in 'suspycous darke and inconvenyent places': young men were being corrupted, were wasting their masters' goods and depriving them of true service and due obedience, and in future acting was to be confined to the houses of noblemen, substantial citizens, the open streets, or the halls of the City companies. The City companies, however, do not seem to have encouraged the performance of plays in their halls after about 1540, perhaps because the drama was becoming more and more secular and more and more commercialized. One of the few later examples is the acting of Mulcaster's boys in Merchant Taylors' hall in 1572, but this was a special case, for the school was the company's and the performance for the pleasure of the master and wardens and the parents. Even so the audience was unruly, and in 1573 all plays in the hall were banned. City prohibitions continue of interludes performed in unsuitable places or on days likely to lead to rioting, and prohibitions of performances on Sundays and holidays are repeated again and again in the forties and fifties.

The earliest evidence of a prohibition in time of plague belongs to 1564, but no doubt it was not the first occasion on which assemblies of all kinds were forbidden on grounds of public health. By reason of this scourge the actors were to lose years of acting time before the theatres were closed altogether in 1642. The recommendation and comment of Bishop Grindal in this year of plague is prophetic of the policy the City was to attempt to enforce: prohibit all plays in the City and within three miles of it 'for one whole yeare (and iff it wer for ever, it werc nott amisse)', for players are 'an idle sorte off people, which have ben infamouse in all goode common weales: I meane these Histriones, common playours.' In the late sixties orders begin to limit performances to afternoon hours, an interesting injunction of January 1569 ordering that houses, inns, and brewhouses used for common plays must not be used for such a purpose after the hour of 5: a month later an injunction limited the playing hours to between 3 and 5. In 1574 the City Remembrancer, Thomas Norton, part-author of *Gorboduc*, in a report to the Lord Mayor reminded him of the danger in plague-time of 'unnecessarie and scarslie honeste resorts to plaies' and added: 'To offend God and honestie is not to cease a plague.' At the same time an act of Common Council gave a comprehensive view of the abuses to which play-acting led: unchaste and seditious speeches, waste of time and money, inveigling of orphans and minors, picking and cutting of purses, frays and quarrels and riots, incontinency, slaughter of the Queen's subjects by faulty structures and the careless use of weapons (as in 1587 during a performance of *Tamburlaine*), and the dangers to public health. The picture is as black as it could be painted. Now players who uttered 'unfytt and uncomelye matter' were threatened with imprisonment, plays were to be censored by the Lord Mayor and court of aldermen, and the houses and inns where plays might be acted were to be licensed.

How far these regulations were operative cannot be known. At least they show that whether from motives moral or prudential or both the City's opposition to the theatres was stiffening. Goaded by the preachers and pamphleteers and their diatribes from 1577 onwards, the City now came out strongly for total suppression. Writing to the Lord Chancellor in 1580 the Lord Mayor called actors 'a very superfluous sort of men' and wished they might be wholly put down. In 1582 the City forbade freemen to allow any servants, apprentices, journeymen, or children to attend plays at any time anywhere: like many another order it worked out admirably on paper but failed when put into practice. The climax was reached in 1584 when the City succeeded in suppressing the Theatre and the Curtain, the players petitioned the Council, and the City in answering the petition observed how 'uncomely' it was for youth to run straight from prayers to plays, from God's service to the Devil's: 'To play in plagetime is to encreasce the plage by infection: to play out of plagetime is to draw the plage by offendinges of God upon occasion of such playes.' The City's language has become indistinguishable from that of Stockwood or Stubbes.

Turn from this welter of evidence, of which it has only been possible to present a summary, to the governmental control of the drama as exercised by the Privy Council. The Council was as anxious as the City to prevent playing in time of plague, to stop rioting, to put down sedition and heresy and immorality, but with a Queen on the throne who considered plays a necessary part of her recreation they could never have agreed to a total prohibition even if they had favoured it. Burghley perhaps may have wished it, and perhaps Knollys, but they were helpless. In 1581 Leicester, known to be a supporter of the stricter sort, was strongly rebuked by John Field, one of the preachers who fulminated against the stage and father-to-be of the actor-dramatist Nathan Field, and was told that if he knew what sinks of sin the players were he would never once look towards them. But he continued to act as patron of a company till his death in 1588.

Evidence of the Council's intervention on behalf of the players is most plentiful and most important between 1572 and 1584 when the City was most hostile. In 1573 and 1574 they commanded the Lord Mayor to admit certain Italian players into the City. Thomas Norton's objection to 'the unchaste, shamelesse and unnaturall tomblinge of the Italion Weomen' as an offence to God and honesty shows the City's point of view. The same privilege was demanded for another Italian company in 1578. In December of that year the Lord Mayor was required to admit the Children of the Chapel and Paul's and the companies of the Lord Chamberlain, Warwick, Leicester, and Essex, for the reason that these companies were to play at Court before the queen at Christmas. The argument that players must be allowed to exercise their craft in order to prepare themselves for court performances was used again and again by the Council and of course by the players. It was in vain for the City to protest, as they did in 1584 and with some truth, that the players could not possibly present before the Queen 'such playes as have ben before commonly played in open stages before all the basest assemblies in London and Middlesex'. Playing in time of Lent was forbidden from 1579, an order which must have pleased the City. In November 1581 the Council had to remind the City that the plague was almost over and players should be allowed to act 'convenient matters for her highnes solace this next Christmas': similar interventions were necessary in 1582 and twice in 1583. In 1581 the Council seems to have acceded to the City's request that there should be no Sunday playing, but there is plenty of evidence that the unruly players did not always observe the ban. After all, the Queen had seen nothing amiss in attending a performance of the *Aulularia* on a Sunday night in King's College Chapel. But that was in 1564.

After 1584 the press campaign against the theatres slackens, if not the pulpit campaign, and the unequal contest between the Council and the City shows signs of abatement. It is significant that in 1583 a company chosen from the best actors in other companies was taken under the patronage of the Queen herself. The Council, while willing to forbid playing on Sundays and in Lent, remained adamant on the general question of toleration. We may remind ourselves that if the theatres had been suppressed we should have had no Elizabethan drama and no Shakespeare, only a surreptitious hole-and-corner affair such as managed to survive between 1642 and the Restoration. By 1584 dramatists and players could realize that they had a friend at court, and proverbially that was worth a penny in the purse. So long as England had a court they had little to fear. When that went, they were doomed.

# CHRONOLOGICAL TABLE

(a.) = acted, ent. = entered in Stationers' Register. (prc.) = preached. (w.) = written.

Otherwise, the date given for tests is the date of publication.

The name of a classical author in capitals means that some or all of his works appeared in their *editio princeps*; if only some, the *editio princeps* of other works is shown by the recurrence of his name with a Roman numeral after it (e.g. EURIPIDES II). N.B. The appearance of a translation is no proof that the *editio princeps* of the original text had already appeared. A title enclosed in square brackets (e.g. Lodge, [*Honest Excuses*]) is one supplied but not found in any old copy. A question-mark between an author's name and a title means doubt about the attribution; elsewhere, doubt about the date.

| Date | Public Events | Private Events |
|---|---|---|
| 1485 | Henry of Richmond crowned as Henry VII. Innocent VIII is Pope. James III K. of Scotland. | Rudolf Agricola d. |
| 1486 | Maximilian I elected K. of the Romans. Cardinal Morton Archbishop of Canterbury. | Plautus, *Menaechmi* acted at Ferrara. |
| 1487 | Lambert Simnel crowned K. of England (in Dublin). Diaz rounds Cape of Good Hope. | |
| 1488 | James III of Scotland murdered; succeeded by James IV. | Duke Humphrey's Library opened in Oxford.<br>Miles Coverdale b. |
| 1489 | | Thomas Cranmer b.<br>*Mémoires* of Philippe de Commines begun (end 1498). |
| 1490 | | Sir Thomas Elyot b.?<br>Sir David Lindsay b. |
| 1491 | | W. Caxton d. |
| 1492 | Spaniards conquer Granada. Lorenzo de' Medici d. Columbus sails from Spain and discovers San Salvador. Pope Alexander VI. | Pietro Aretino b. |
| 1493 | Maximilian I succeeds to Empire. Columbus discovers Jamaica. | |
| 1494 | Charles VIII invades Italy. | Hans Memlinc d.<br>G. Pico della Mirandola d.<br>Boiardo d.<br>Poliziano d. |
| 1495 | | Leonardo da Vinci's *Last Supper*.<br>John Bale b.<br>Rabelais b.? |
| 1496 | | Colet begins to lecture in Oxford. |
| 1497 | Vasco da Gama rounds Cape. | John Heywood b.<br>Clement Marot b. |

| *English and Scotch Texts* | *Greek, Latin, and Continental Vernaculars* |
|---|---|
| Caxton's edn. of Malory's *Morte Darthur*. | |
| *The Boke of St. Albans*. | PROBUS.<br>VITRUVIUS?<br>BATRACHOMYOMACHIA. |
| | VEGETIUS.<br>AELIAN.<br>FRONTINUS. |
| | HOMER. |
| | Poliziano, *Miscellanea*.<br>Villon's *Grand Testament* and *Petit Testament*. |
| Caxton, *Eneydos*.<br>Caxton, *Art and Craft to know well to die*. | |
| | ? THEOCRITUS I.<br>ISOCRATES.<br>? HESIOD I. |
| Walter Hylton, *Scala Perfectionis*. | Sebastian Brant, *Narrenschiff*.<br>First Aldine publication.<br>QUINTILIAN III.<br>ANTHOLOGIA GRAECA.<br>? MUSAEUS. |
| John Trevisa, trans. of *De Proprietatibus Rerum* of Bartholomaeus Anglicus. | Boiardo, *Orlando Innamorato*.<br>? EURIPIDES I.<br>CALLIMACHUS I.<br>Terence, *Comoediae* (Pynson) |
| *Mandeville's Travels*. | THEOPHRASTUS I.<br>APOLLONIUS RHODIUS.<br>ARISTOTLE I.<br>THEOGNIS.<br>THEOCRITUS I.<br>LUCIAN.<br>BION.<br>MOSCHUS.<br>HESIOD. |
| | ZENOBIUS. |

| *Date* | *Public Events* | *Private Events* |
|---|---|---|
| 1498 | Machiavelli appointed Florentine Secretary. Cabot discovers Labrador. Vasco da Gama reaches Calicut. Columbus reaches S. American mainland. | Erasmus settles at Oxford. |
| 1499 | Perkin Warbeck and Earl of Warwick beheaded. Swiss independence agreed. | |
| 1500 | | R. Henryson d.?<br>G. della Casa b.<br>Geo. Cavendish b.<br>Benvenuto Cellini b.<br>Erasmus leaves England. |
| 1501 | P. Arthur m. Catherine of Aragon. | |
| 1502 | P. Arthur d. | |
| 1503 | Alexander VI d.; Pius III succ. and d.; Julius II succ.; James IV m. P. Margaret. P. Henry betrothed to Catherine of Aragon. Q. Elizabeth d. | Pontanus d.<br>Wyatt b.<br>Paston letters come to an end (begun *c.* 1425). |
| 1504 | | Matt. Parker b.<br>Colet, Dean of St. Paul's. |
| 1505 | | Knox b.<br>Nicholas Udall b.?<br>Erasmus returns to England. |
| 1506 | Treaty with Burgundy | Columbus d.<br>Geo. Buchanan b.<br>Leland b.?<br>Erasmus leaves England. |
| 1507 | | Sturmius b.<br>Chepman and Myllar's press set up.<br>Abbot Damian's attempt at aviation. |
| 1508 | | Ariosto, *La Cassaria* (a.) |

| *English and Scotch Texts* | *Greek, Latin, and Continental Vernaculars* |
|---|---|
| Lydgate, *Assembly of the Gods.* | APICIUS.<br>'PHALARIS'.<br>ARISTOPHANES I. |
| | ETYMOLOGIUM MAGNUM.<br>SUIDAS. |
| Skelton, *Bowge of Court.*<br>*Sir Bevis of Hampton.*<br>*Guy of Warwick* (may be 2nd edn.)<br>*Everyman* tr. from Dutch? | Erasmus, *Adagia.*<br>'ORPHEUS'.<br>*La Celestina?* |
| Douglas, *Palace of Honour* (w.). | Abrabanel (= Leone Ebreo) *Dialoghi di Amore* (w.).<br>Sannazaro, *Arcadia.* |
| Richard Arnold, *Chronicle.* Skelton, *Speculum Principis* (w.). | HERODOTUS.<br>SOPHOCLES.<br>THUCYDIDES. |
| Wm. Atkinson, tr. *The Imitation of Christ.*<br>Dunbar, *Thistle and Rose* (w.).<br>Pseudo-Barclay, *Castle of Labour* (Paris)? | EURIPIDES.<br>Erasmus, *Enchiridion.*<br>Pellicanus, Hebrew Grammar. |
| Skelton, *Philip Sparrow* (w.)? | Sannazaro, authorized edn. of *Arcadia.*<br>Erasmus' Latin translation of Euripides, *Hecuba* and *Iphigenia.* |
| Publication of fourteenth-century *Contemplations of the dread and love of God* (R. Rolle). | Boiardo, first extant edn. of *Orlando Innamorato.*<br>Reuchlin, Hebrew Grammar and Dictionary. |
| Anon., *Kind Kittock.* Dunbar, *Ballade of Ld. B. Stewart*; *Flyting of D. and Kennedy*; *Golden Targe*; *Two Marryit Women, &c.*; *Lament for Makaris*; *Dance of the VII . . Sins*; *Thistle and Rose.* Fisher, *Treatise concerning the Fruitful Sayings of David in the VII Penitential Psalms.* | |

| Date | Public Events | Private Events |
| --- | --- | --- |
| 1509 | Acc. Henry VIII; m. Catherine of Aragon. | Calvin b.<br>Telesius b.<br>Erasmus returns to England.<br>Ariosto, *I Suppositi* (a.) |
| 1510 | | Colet founds St. Paul's School.<br>Botticelli d. |
| 1511 | Henry joins the Holy League. Anglo-Scotch tension. | Colet's Sermon to Convocation (prc.).<br>Erasmus becomes Greek Reader at Cambridge. |
| 1512 | | Mercator b. |
| 1513 | Leo X succ. Henry invades France. Battle of Flodden. James IV killed; James V acc. | Mantuan d.<br>Thos. Smith b.<br>Bibbiena, *La Calandria* (a.) |
| 1514 | Wolsey becomes Archb. of York. | Cheke b.<br>Poynet b.<br>Vesalius b. |
| 1515 | Wolsey becomes Cardinal. Act against Enclosures. | Ascham b.<br>Ramus b.<br>Wierus b. |
| 1516 | | Hieron. Bosch d.<br>Trithemius d.<br>Skelton's *Magnificence* (a.)? |
| 1517 | Luther puts forward his Wittenberg Theses. Wolsey's Commission (on Enclosures). | Thos. Cooper b.?<br>Jn. Foxe b.<br>Surrey b. |
| 1518 | Wolsey becomes *Legatus a latere.* Treaty of Universal Peace. | Robt. Crowley b.<br>Ninian Winzet b. |
| 1519 | Cortes invades Mexico. Magellan begins his voyage round the world. | Colet d.<br>Leonardo da Vinci d.<br>Beza b.<br>Grimald b.? |

| *English and Scotch Texts* | *Greek, Latin, and Continental Vernaculars* |
|---|---|
| Pseudo-Chaucer, *The Maying or Disport of Chaucer*. Publication of *Remedy against . . . Temptation* (R. Rolle). | |
| Barclay, *Ship of Fools*. Copland, *King Appolyn*. Fisher, *Month's Mind of P. Margaret*. Funeral Sermon on Henry VII. Hawes, *Pastime of Pleasure*; *Conversion of Swearers*; *Joyful Meditation*. | Erasmus, *Moriae Encomium*.<br>PLUTARCH I.<br>Macropedius, *Asotus* (w.) |
| Anon., *Cock Lorell's Boat*. More, *Merry Jest?*; *Life of Pico?* | Agrippa, *De Occulta Philosophia* (w.). |
| Copland, *Helyas*. | |
| Anon., *Flowers of Ovid*. Bradshaw, *Life of St. Werburge* (w.). Douglas, tr. of *Aeneid* (w.). Skelton, *Ballad of Scottish King*. | PINDAR.<br>PLATO.<br>Machiavelli, *Il Principe* (w.). |
| | Complutensian Polyglot N.T.<br>Quintianus Stoa, *Theocrisis*. |
| Barclay, *Eclogues*, I–III? | TACITUS.<br>*Epistolae Obscurorum Virorum*.<br>Trissino, *Sophonisba*. |
| Fabyan, *Chronicle*. | Ariosto, *Orlando Furioso*.<br>Erasmus, *Novum Instrumentum*.<br>More, *De optimo reipublicae statu deque nova insula Vtopia* (Louvain).<br>Pomponatius, *De Immortalitate*.<br>Paulus Ricius, Lat. tr. of Cabbalistic *De Portis Vitae*. |
| | Complutensian Polyglot Bible.<br>Reuchlin, *De Arte Cabbalistica*. |
| | ÆSCHYLUS.<br>Luther, *Resolutiones*. |
| | Reuchlin, *De Recta Latini Graecique Sermonis Pronuntiatione Dialogus*. |

| *Date* | *Public Events* | *Private Events* |
|---|---|---|
| 1520 | Field of the Cloth of Gold. Henry entitled *Fidei Defensor* by the Pope. | Raphael d.<br>Churchyard b.<br>Machiavelli, *Mandragola* (a.)? |
| 1521 | Diet of Worms. Papal ban on Luther. | Fisher's sermon against Luther (prc.).<br>Dunbar d.?<br>Richard Eden b.? |
| 1522 | Adrian VI succ. Charles V visits England. | Abrabanel d.?<br>Douglas d.<br>Reuchlin d.<br>Du Bellay b.<br>Jewel b. |
| 1523 | Clement VII succ. | Hawes d.? |
| 1524 | | Linacre d.<br>Richard Edwards b.<br>Camoens b.<br>Ronsard b.<br>Thos. Tusser b.? |
| 1525 | Battle of Pavia. | Palestrina b.<br>Jn. Stow b.?<br>Thos. Wilson b.?<br>Machiavelli, *Clizia* (a.) |
| 1526 | Pizarro conquers Peru. Turks capture Buda. Prohibition and burning of English N.T.s. Persecution of English Protestants (1526–7). | Jn. Lesley b. |
| 1527 | Sack of Rome. | Machiavelli d.<br>Ortelius b. |
| 1528 | Wolsey suppresses some smaller religious houses. | Dürer d.<br>Ruzzante, *Bilora* (a.)?<br>Ariosto, *Negromante* and *La Lena* (a.). |
| 1529 | Synod of Örebro; Lutheranism established in Sweden. Fall of Wolsey. Rise of Thos. Cromwell. Henry meets Cranmer. More becomes Chancellor. The 'Reformation Parliament'. | Castiglione d.<br>Skelton d. |

| *English and Scotch Texts* | *Greek, Latin, and Continental Vernaculars* |
| --- | --- |
| Barclay, tr. of Sallust's *Jugurtha*?<br>Murdoch Nisbet, Scotch N.T.?<br>Anon., *Youth* (w.)? | Henry VIII, *Assertio VII Sacramentorum.*<br>Luther, *Von der Freiheit.*<br>Pomponatius, *De Admirandorum Effectuum Causis* (w.). |
| Barclay, *Introductory to . . . French.* Bradshaw, *Life of St. Werburge.* Bradshaw (?), *Life of St. Radegunde*? Fisher, *Sermon against Luther.* Skelton, *Speak Parrot* (w.). | Ariosto, *Orlando Furioso* (rev. edn.). |
| Anon., *The World and the Child.* | Luther, German N.T., and *Contra Henricum Regem Angliae.*<br>Vives, edn. of St. Augustine's *De Civitate Dei.* |
| Barclay, *Mirror of Good Manners.* Berners, tr. of Froissart, vol. i. Skelton, *Garland of Laurel.* | |
| Leonard Cox, *Art or Craft of Rhetoric.* | |
| Berners, Froissart, vol. ii. Tyndale, Cologne N.T. | Bembo, *Prose della Volgar Lingua.*<br>Münster, Hebrew Grammar.<br>Trissino begins *Italia Liberata* (blank verse). |
| Anon., *Hundred Merry Tales.* Tyndale, Worms N.T., *Prologue to Romans.* | Hector Boece, *Historia Scotorum.*<br>F. de Oviedo, *Summario.* |
| Anon., *Calisto and Melibea*?<br>*Gentleness and Nobility*? | Colet, *Aeditio.*<br>Vida, *De Arte Poetica.* |
| Anon., tr. *Jest of . . . Howleglass.* Fisher, *Sermon . . . concerning certain Heretics.* More, *Dialogue with the Messenger.* Wm. Roy, *Read me and be not Wroth* (Strassburg). Tyndale, *Wicked Mammon*; *Obedience of a Christian Man.* | Castiglione, *Il Cortegiano.*<br>Pagninus, Latin O.T. |
| Simon Fish, *Supplication for Beggars.* More, *Supplication of Souls.* Jn. Rastell, *The Pastime of People.* | Zürich Bible.<br>Paracelsus, *Practica.*<br>Fullonius, *Acolastus.*<br>Barthélémy, *Christus Xylonicus.* |

| Date | Public Events | Private Events |
|---|---|---|
| 1530 | Diet of Augsburg. Threatened with *Praemunire*, the English clergy give Henry £100,000 and call him Supreme Head of the Church 'as far as the law of Christ allows'. | S. Fish d.<br>Wolsey d.<br>Sannazaro d.<br>Bodinus b.<br>Thos. Hoby b.<br>Richd. Mulcaster b.?<br>Alexander Scott b.?<br>Jn. Whitgift b.? |
| 1531 | Act against vagabonds. Henry forbids the burning of Edw. Crome because his opinions include a denial of Papal Supremacy. | Zwinglius d.<br>*Gl'Ingannati* (a.). |
| 1532 | Henry divorces Catherine of A. Convocation makes drastic concessions to Henry. More resigns. Annates Act. | Wm. Allen b.<br>Jn. Hawkins b.<br>Thos. Norton b. |
| 1533 | Cranmer becomes Archb. of Canterbury. Henry excommunicated. m. Anne Boleyn. Act in Restraint of Appeals. Arrest of Eliz. Barton (the 'Maid of Kent'). | Ariosto d.<br>Berners d.<br>Jn. Frith burnt.<br>Montaigne b. |
| 1534 | Paul III succ.<br>Acts of Succession and Supremacy. | Correggio d.<br>Jn. Leland touring England.<br>W. Rastell ceases to print. |
| 1535 | Henry, 'supreme Head on earth', Cromwell's visitation of Monasteries. More and Fisher executed. | Cornelius Agrippa d.<br>Thos. Cartwright b.<br>Thos. Legge b.<br>Thos. North b. |
| 1536 | Anne Boleyn's miscarriage, and execution. Henry m. Jane Seymour. New Act of Succession. Act against vagabonds. X Articles. Beginning of Pilgrimage of Grace. Smaller abbeys suppressed. | Erasmus d.<br>Jn. Rastell d.<br>Tyndale strangled.<br>Arthur Golding b.<br>Thomas Sackville b. |
| 1537 | The Pope condemns slavery. Beginning of Calvin's Theocracy at Geneva; banishment of Caroli; sumptuary legislation. P. Edward b. Jane Seymour d. | *Thersites* (a)? |

| *English and Scotch Texts* | *Greek, Latin, and Continental Vernaculars* |
| --- | --- |
| Colet, tr. of his *Sermon in Convocation.* Copland, *High Way to the Spital House*? Lindsay, *Complaint and Testament of the Papyngo* (w.). Jn. Rastell, *New Book of Purgatory.* Roy, *A Proper Dialogue* (Marburg). Tyndale, *Practice of Prelates*; tr. *Pentateuch. Terence in English*? Redford, *Wit and Science*? | Fracastorius, *Syphilis sive Morbus Gallicus.*<br>Textor, *Dialogi.* |
| Elyot, *Book of the Governor.* Tyndale, *Answer to More*; tr. Jonah. | PARTHENIUS.<br>Cornelius Agrippa, *De Occulta Philosophia* and *De Incertitudine et Vanitate Scientiarum.*<br>Vives, *De Disciplinis.* |
| Wm. Thynne's edn. of Chaucer. Elyot, *Pasquil the Plain.* G. Hervet, tr. of Xenophon's *Oeconomicus.* More, *Confutation of Tyndale's Answer, Letter* (against Frith). | Ariosto, *Orlando Furioso* (rev. edn.).<br>Erasmus, *Apophthegmata.*<br>Rabelais, *Les Horribles faits du roy Pantagruel.* Sixt Birck, *Susanna.* |
| Jn. Gaw, *Right Way to the Kingdom of Heaven.* Elyot, *Of the Knowledge which Maketh a Wise Man.* More, *Answer to a Poisoned Book*; *Apology*; *Confutation*, Pt. II; *Debellation.* N. Udall, *Flowers for Latin Speaking.* | |
| Berners, tr. *Huon of Bordeaux.* Colet, *Right Fruitful Admonition.* Elyot, *Castle of Health*; tr. Sermon of St. Cyprian; tr. 'Rules of Christian Life' from Pico; tr. 'Doctrine of Princes' from Isocrates. More, *Dialogue of Comfort* (w.). Thos. Starkey, *Dialogue* (w.). R. Whytington, tr. 3 bks. of Cicero's *De Officiis.* | Lefèvre, French Bible.<br>Münster, Latin O.T.<br>Polydore Vergil, *Historia Anglica.*<br>Rabelais, *La Vie du grand Gargantua.* |
| Berners, *Golden Book of Marcus Aurelius.* Fisher, *Spiritual Consolation to his Sister.* Copland, *Complaint of them that be too soon married*; *Complaint of them that be too late married*? Coverdale, *Bible out of Dutch and Latin* (Zürich?); *Biblia the Bible.* | EPICTETUS.<br>Abrabanel, *Dialoghi di Amore.*<br>Olivetan's French Bible.<br>Vida, *Christias.*<br>Macropedius, *Rebelles.* |
| Pseudo-Cheke, *Remedy for Sedition.* | Calvin, *Institutio.* |
| Cranmer, *Institution of a Christian Man* (= 'The Bishop's Book').<br>Lindsay, *Deploration of the Death of Q. Magdalene* (w.). 'Matthew's' Bible. | Jn. Fortescue, *De Laudibus Legum Angliae.*<br>Latimer, *Concio ante Inchoationem Parlamenti.* |

| *Date* | *Public Events* | *Private Events* |
|---|---|---|
| 1538 | James V of Scotland m. Marie de Guise. In England the English Bible ordered to be placed in every church. Persecution of Anabaptists. Breaking of images. Execution of Henry Pole. | Thos. Starkey d.<br>Guarini b.<br>Reg. Scott b.? |
| 1539 | Greater abbeys suppressed. Statute of VI Articles (whereby heresy becomes a felony at Common Law). Act of Proclamations. | Geof. Fenton b.?<br>Humph. Gilbert b.?<br>Geo. Gascoigne b.? |
| 1540 | Institution of the Jesuits. Calvin, after banishment, returns to Geneva. Fall and execution of Cromwell. Henry m. Anne of Cleves; marriage annulled; he m. Catherine Howard. | Vives d.<br>Barnaby Googe b.<br>Wm. Painter b.<br>Geo. Turberville b.?<br>Lindsay's *Satire of Three Estates* (a.). |
| 1541 | Henry assumes title 'King of Ireland' and 'Head of the Church in Ireland'. | Paracelsus d.<br>Giraldi, *Orbecche* (a.). |
| 1542 | Establishment of the Roman Inquisition. Henry invades Scotland without declaring war and executes Catherine Howard. Act against vagabonds. James V of Scotland d.; his daughter Mary b. and acc. | Wyatt d.<br>Bellarmine b.<br>Barnaby Rich b.? |
| 1543 | Mary Q. of Scots betrothed to the Dauphin; goes to France. Henry m. Catherine Parr. Act 'for the Advancement of True Religion' (restricts Bible reading and condemns unauthorized translations). | Copernicus d.<br>Thos. Deloney b.?<br>Du Bartas b.<br>Baptista Porta b.<br>Textor dialogue acted at Cambridge. |
| 1544 | Hertford's invasion of Scotland. War with France; capture of Boulogne. | Marot d.<br>Tasso b.<br>Geo. Whetstone b.?<br>Buchanan's translations of Euripides, *Medea* and *Alcestis* (a.)? |
| 1545 | Council of Trent opens. Confiscation of Chantries. | Thos. Bodley b.<br>Gabr. Harvey b.? |
| 1546 | Murder of Cardinal Beaton. Peace with France. | Brinkelow d.<br>Elyot d.<br>Luther d.<br>Desportes b. |

| *English and Scotch Texts* | *Greek, Latin, and Continental Vernaculars* |
|---|---|
| Elyot, *Latin-English Dictionary.* Lindsay, *Complaint of the Papyngo.* Bale, *King John* (first version), *Temptation of our Lord, Three Laws, John Baptistes.* | |
| The 'Great' Bible. Richd. Taverner's Bible. Elyot, *Banquet of Sapience.* | DIODORUS SICULUS.<br>Joachim Rheticus (a pupil of Copernicus), *Narratio prima de Revolutionibus*, first publication of the Copernican theory. |
| Berners, tr. of San Pedro's *Castle of Love.* Douglas, *Palace of Honour* ? (Edinburgh). Elyot, tr. of Eucolpius, *Image of Governance.* Palsgrave, *Ekphrasis of Acolastus.* | Calvin, French version of *Institutio.*<br>*Comoediae et Tragoediae* (printed by Brylinger). |
| Brinkelow, *Lamentation of a Christian against London.* Edw. Hall, *Union of the Families of Lancaster and York* ? (undated fragment). Wm. Lyly, *Introduction to the VIII Parts of Speech.* Robt. Record, *Ground of Arts teaching Arithmetic.* N. Udall, tr. Erasmus' *Apophthegmata.* | |
| Composite, *The Necessary Doctrine and Erudition of any Christian Man* (= 'The King's Book'). Richd. Grafton, *Chronicle* (i.e. Jn. Harding's with continuation which includes More's *Life of Richard III*). | Copernicus, *De Revolutionibus.*<br>Ramus, *Aristotelicae Animadversiones* and *Dialecticae Partitiones.*<br>Grimald, *Christus Redivivus.* |
| Bale, *Brief chronicle concerning Sir John Oldcastle, Lord Cobham* (Antwerp?). Heywood? *The Four PP.* | Leland, *Assertio . . . Arturii Regis.* |
| Ascham, *Toxophilus.* Churchyard, *Mirror of Man*? Elyot, *Defence of Good Women.* 'King's *Primer*' (including Cranmer's Litany). Hugh Rhodes, *Book of Nature* ? Skelton, *Certain books compiled by* ? | EURIPIDES II.<br>Rabelais, *Pantagruel* iii. |
| Bale, *Acts of English Votaries*; *First Examination of Anne Askew.* Jn. Heywood, first *Dialogue of Proverbs.* | Fracastorius, *De Contagione.* |

| *Date* | *Public Events* | *Private Events* |
|---|---|---|
| 1547 | Henry VIII d.; Edward VI succ. Execution of Surrey. Scots defeated at Pinkie. Somerset becomes Protector. Act of Proclamations repealed. Many laws against heresy abrogated. Penal slavery for some types of vagabonds. | Bembo d.<br>R. Copland d.<br>Ed. Hall d.<br>Cervantes b.<br>Lipsius b.<br>Richd. Stanyhurst b. |
| 1548 | Somerset's Commission on Enclosures. Arrest of Gardiner. | Giordano Bruno b.<br>Geo. Peele b.<br>Geo. Pettie b. |
| 1549 | Paul III d. English retreat in Scotland; the war becomes desultory. War with France. Kett's rebellion. Act of Uniformity. | Giles Fletcher (senior) b.?<br>Martin Bucer arrives in England. |
| 1550 | Julius III succ. Treaty of Boulogne. Removal of R.C. bishops. Bishops' incomes docked. New Ordinal. Abolition of certain ceremonies. Burning of books at Oxford. | Trissino d.<br>Robt. Browne b.?<br>Jn. Napier, inventor of logarithms, b.<br>Henry Smith b.?<br>Beza, *Abraham Sacrifiant* (a.). |
| 1551 | Council of Trent, Session II. Bolsec banished from Geneva for denying Predestination. Treaty of Angers. | Bucer d.<br>Nich. Breton b.?<br>Wm. Camden b. |

| *English and Scotch Texts* | *Greek, Latin, and Continental Vernaculars* |
|---|---|
| Wm. Baldwin, *Treatise of Philosophy.* Bale, *The Latter Examination of A. Askew.* Cranmer (with Bonner, Grindal, Harpsfield, and others), *Certain Sermons or Homilies.* Thos. Sternhold and Jn. Hopkins, first book of Psalms. | Amyot, Fr. tr. of Heliodorus. Ramus, *Institutiones Dialecticae.* Trissino, *Italia Liberata.* *Comoediae et Tragoediae* (printed by Oporinus). Dolce, *Dido.* |
| Brinkelow, *Complaint of Roderick Mors.* R. Crowley, *Information and Petition against the Oppressors of the Poor Commons.* Hall, *Union of . . . York and Lancaster,* 2nd edn. Latimer, *Notable Sermon in the Shrouds* (= Plough Sermon), Lindsay, *The Tragedy of the . . . Cardinal,* N. Udall, tr. Erasmus, Paraphrase of N.T. | Bale, *Illustrium Majoris Britanniae Scriptorum Summarium.* Foxe, *De Non Plectendis Morte Adulteris.* Rabelais, *Pantagruel* iv. Ramus, *Scholae Dialecticae.* Grimald, *Archipropheta* |
| Anon., *The Complaint of Scotland* (w.). Baldwin, *Canticles or ballads of Solomon . . . in English metres. Book of Common Prayer.* Thos. Challoner, tr. Erasmus, *Praise of Folly.* Cheke, *The Hurt of Sedition.* Cooper-Lanquet, *Epitome of Chronicles* (Thos. Lanquet, completed by Cooper). Crowley, *The Voice of the Last Trumpet.* Latimer, *The First Sermon before the King's Majesty.* Leland–Bale, *The Laborious Journey and Search of John Leland . . . a New Year's Gift . . . enlarged by J. Bale.* Jn. Poynet, *Tragedy or Dialogue of the Bishop of Rome* (tr. from B. Ochino). Sternhold and Hopkins, second book of Psalms. Wyatt, *Certain Psalms drawn into English Metre.* | Du Bellay, *Défense et Illustration de la langue française* and *Olive.* |
| E.A., *Notable Epistle of Dr. Gribaldi.* Cranmer, *Defence of the true Doctrine of the Sacrament.* Crowley, *One and Thirty Epigrams, Way of Wealth.* Jn. Harington, tr. Cicero's *De Amicitia.* Jn. Heywood, *Dialogue of Proverbs* (enlarged). Wm. Hunnis, *Certain Psalms in English Metre.* Latimer, *A most faithful Sermon . . . before the King's . . . Majesty, A Sermon . . . at Stamford.* Thos. Nicolls, tr. of Thucydides. Richd. Sherry, *Treatise of Schemes and Tropes.* | Ramusio, *Navigationi.* Ronsard, *Odes* i–iv. Muretus, *Julius Caesar.* |
| Cranmer, *An Answer to Stephen Gardiner.* Crowley, *Pleasure and Pain; Fable of Philargyrie.* Ralph Robinson, tr. of More's *Utopia.* Thomas Wilson, *Rule of Reason.* | Cardan, *De Subtilitate Rerum.* Pollanus, *Liturgia Sacra.* |

| Date | Public Events | Private Events |
|---|---|---|
| 1552 | Execution of Somerset. 2nd Act of Uniformity. | Barclay d.<br>Philemon Holland b.<br>Ralegh b.?<br>Spenser b.<br>Jodelle, *Cléopâtre Captive* (a.). |
| 1553 | Servetus burnt at Geneva. Edward VI d. Acc. Mary. Sir Hugh Willoughby's voyage begins. | Fracastorius d.<br>Rabelais d.<br>Richd. Hakluyt b.<br>Anthony Munday b.<br>N. Udall's *Roister Doister* (a.)?<br>*Respublica* (a.). |
| 1554 | Rebellion of Sir Thos. Wyatt. Execution of Lady Jane Grey. Mary m. Philip of Spain. England reconciled with Rome but Mary retains title of 'Supreme Head'. Church lands not restored. Parliament refuses to revive the old Heresy Laws. | Stephen Gosson b.<br>Fulke Greville b.<br>Richd. Hooker b.<br>Jn. Lyly b.?<br>Philip Sidney b.<br>Edinburgh performance of Lindsay's *Satire of the Three Estates*. |
| 1555 | Paul IV succ. Agreement at Augsburg establishing toleration of Lutheranism. Persecution of Protestants in England; Latimer and Ridley burnt. Muscovy Company founded. | Polydore Vergil d.<br>Lindsay d.<br>Lancelot Andrewes b.<br>Malherbe b.<br>W. Gager b. |
| 1556 | Philip II now King of Spain, the Indies, Netherlands, Naples, &c.; is excommunicated. Cranmer burnt. | Aretino d.<br>G. della Casa d.<br>Poynet d.<br>Nich. Udall d.<br>Alex. Montgomery b.?<br>Peele b. |
| 1557 | War with France. | Stationers' Company incorporated.<br>Thos. Kyd b.?<br>Thos. Watson b. |
| 1558 | Loss of Calais. Mary d. Elizabeth I acc. | Olaus Magnus d.<br>J. C. Scaliger d.<br>Robt. Greene b.<br>Thos. Lodge b.<br>Wm. Perkins b.<br>Wm. Warner b.? |

| *English and Scotch Texts* | *Greek, Latin, and Continental Vernaculars* |
| --- | --- |
| *Book of Common Prayer* II. | Mercator, *De Usu Annuli Astronomici.*<br>Gomára, *Istoria de las Indias.*<br>Ronsard, *Amours de Cassandre.* |
| Douglas, *Aeneid* (corrupt); *Palace of Honour* (London). Eden, tr. of Sebastian Münster, *Treatise of the New India.* More, *Dialogue of Comfort.* Wilson, *Art of Rhetoric.* | *Lazarillo de Tormes.*<br>Ronsard, *Odes.* |
| Lindsay, *Dialogue betwixt Experience and Ane Courtier.* | *Ποιμάνδρης* (Hermetic writings).<br>Lat. tr. of Achilles Tatius.<br>Bandello, *Novelle* i–iii.<br>Bellus, *De Haereticis* (w.).<br>Foxe, *Commentarii.*<br>Palestrina, First Book of Masses. |
| Anon., *Institution of a Gentleman.* Berners, *Arthur of Little Britain* ? Eden, tr. of *Decades of Peter Martyr.* Heywood, *Two Hundred Epigrams. Mirror for Magistrates* I (Wm. Baldwin's *Fall of Princes,* &c., extant only in a fragment). | Cheke, *De Pronuntiatione.*<br>Ramus, Fr. tr. *Dialectique.*<br>Olaus Magnus, *De Gentibus Septentrionalibus.*<br>Ronsard, *Amours de Marie, Hymnes.* |
| Nich. Harpsfield, *Life of More* (w.). Heywood, *The Spider and the Fly.* Poynet, *Short Treatise of Politic Power* (Strassburg ?). Genevan metrical *Psalter* I. | |
| Anon., *Courte of Venus* (ent. 57/58); *Sackful of News* (ent. 57/58). Geo. Cavendish, *Life of Wolsey* (w.). More, *English Works.* North, tr. of Guevara, *Dial of Princes.* Surrey, *Certain Books of Virgil's Aeneis.* Tottel, *Songs and Sonnets* (= 'Tottel's Miscellany'). Tusser, *Hundred Good Points.* | Bale, *Catalogus.*<br>Cardan, *De Varietate Rerum.* |
| Knox, *Appellation*; *First Blast of the Trumpet.* Wm. Painter, *Murder of Solyman* ? Thos. Phaer, tr. *Seven First Books of the Aeneidos.* | Du Bellay, *Antiquités de Rome, Poemata.*<br>Marguerite de Navarre, *Heptameron* I. |

| *Date* | *Public Events* | *Private Events* |
|---|---|---|
| 1559 | Pius IV succ. Bull *Cum ex apostolatus*, by which all princes who support Henry forfeit their office. Mary of Scotland m. the Dauphin. Knox revisits Scotland: revolution begins, assisted by Elizabeth. Peace between England and France. M. Parker Archb. of Canterbury. Acts of Uniformity (entailing a fine for not attending church) and of Supremacy (refused by nearly all the Marian bishops and accepted by nearly all the inferior clergy). Ecclesiastical Court of High Commission. Popular iconoclasm and looting of churches. Vestiarian controversy begins. | Geo. Chapman b.? |
| 1560 | Peace with Scotland. Parpaglia sent by the Pope to Elizabeth, is intercepted by Philip of Spain. Statute against the destruction of church monuments. | Melanchthon d.<br>Hen. Chettle b.?<br>Alex. Hume b.?<br>First professional theatre building in Spain. |
| 1561 | English occupation of Le Havre. O'Neill's rebellion in Ireland. | Cavendish d.<br>Francis Bacon b.<br>Jn. Harington b.?<br>Edwin Sandys b.<br>Robt. Southwell b.<br>*Gorboduc* (a.). |
| 1562 | Council of Trent, Session III. French Civil War. Elizabeth's secret treaty of Richmond with the Huguenots. English force at Havre. Hawkins's voyage begun. | Bandello d.<br>Montemayor d.<br>Constable b.<br>Lope de Vega b. |
| 1563 | English refugees unsuccessfully petition Council of Trent for Elizabeth's excommunication. Plague at Havre; Havre evacuated. Plague in London. | Bale d.<br>Arthur Brooke d.<br>Sam. Daniel b.?<br>Drayton b. |

| *English and Scotch Texts* | *Greek, Latin, and Continental Vernaculars* |
|---|---|
| *Book of Common Prayer* III. Bullein, *Government of Health.* Crowley–Cooper–Lanquet chronicle (see 1549) with continuation. Jasper Heywood, tr. Seneca's *Troas. Mirror for Magistrates* II (first extant edn.). | Amyot, F. tr. of Plutarch, Longus.<br>Beza, *De Haereticis a civili magistratu Puniendis.*<br>Du Bellay, *Regrets.*<br>Foxe, *Rerum in Ecclesia Gestarum.*<br>Minturno, *De Poeta.* |
| W. Barkar, tr. of Xenophon's *Cyropaedia.* Geneva Bible. C. T., tr. of Boccaccio, *Pleasant History of Galesus.* Copland, *Jyll of Brentford.* Jasper Heywood, tr. Seneca's *Thyestes.* Jn. Heywood, *Fourth Hundred of Epigrams.* Jewel, *Sermon at Paul's Cross.* Knox, *An Answer to . . . blasphemous Cavillations.* Jn. Rolland, *Seven Sages* (w.?). T. H., *Fable of Narcissus.* R. Wever, *Lusty Juventus* (ent.). | |
| Jn. Awdeley, *Fraternity of Vagabonds* (ent.). Jn. Dolman, tr. of Cicero's *Tusculans.* Eden, tr. Cortes, *Art of Navigation.* Googe, tr. Palingenius, *Zodiac of Life* i–vi. Jasper Heywood, tr. Seneca, *Hercules Furens.* Hoby, tr. Castiglione, *Book of the Courtier.* Thos. Norton, tr. Calvin's *Institution.* Stow, edn. of Chaucer. | Scaliger, *Poetices.* |
| Arthur Brooke, *Romeus and Juliet.* Bullein, *Bulwark against all Sickness.* Cooper, *Answer against the Apology of Private Mass.* Jn. Heywood, *Works.* Jewel, *Apology of Private Mass.* Latimer, *Twenty Seven Sermons.* Sternhold and Hopkins, &c., *The Whole Book* (of Metrical Psalms). Ninian Winzet, *Certain Tractates, The Last Blast of the Trumpet.* | Jewel, *Apologia pro Ecclesia Anglicana.*<br>Rabelais, *Pantagruel* v. |
| Foxe, *Acts and Monuments* (= 'Foxe's Book of Martyrs'). Googe, *Eclogues, Epitaphs, and Sonnets.* Grafton, *Abridgement of Chronicles. Mirror for Magistrates* | Minturno, *Arte Poetica.*<br>More, *Lucubrationes.*<br>Roillet, *Philanire.* |

| Date | Public Events | Private Events |
|---|---|---|
| | XXXIX Articles. Act forbidding conversion of arable land to pasture. Elizabeth's first Poor Law. | Joshua Sylvester b. |
| 1564 | Treaty of Troyes. Influx of Flemish Protestant refugees. | Calvin d.<br>Michelangelo d.<br>Galileo b.<br>Marlowe b.<br>Shakespeare b.<br>Edwards, *Damon and Pithias* (a.).<br>Q. Elizabeth visits Cambridge. |
| 1565 | Attempts to check English and French piracy in the Channel. | Francis Meres b. |
| 1566 | Pius V succ. James (afterwards VI and I) b. | Hoby d.<br>Vida d.<br>Gascoigne's *Supposes* and *Jocasta* (a.).<br>Q. Elizabeth visits Oxford: Edwards, *Palamon and Arcite* (a.). *Sapientia Salomonis* (a.). |
| 1567 | Revolt of the Netherlands. | Thos. Campion b.<br>Anthony Copley b.<br>Monteverdi b.<br>Thos. Nashe b.<br>Pickering, *Horestes* (a.)? *The Marriage of Wit and Science* (a.)? |
| 1568 | English College of Douai founded. Mary Q. of Scots flies to England. Elizabeth seizes treasure on Spanish fleet driven into Plymouth by pirates. Growing tension with Spain. | Ascham d.<br>Coverdale d.<br>Olaus Magnus d.<br>Campanella b.<br>Gervase Markham b.?<br>Hen. Wotton b. |

| *English and Scotch Texts* | *Greek, Latin, and Continental Vernaculars* |
|---|---|
| III (containing the work of Dolman and Sackville). Alex. Neville, tr. of Seneca's *Oedipus*. Richd. Rainold, *Foundation of Rhetoric*. Winzet, *Book of Fourscore Questions* (Antwerp). | |
| Anne Bacon, tr. of Jewel's *Apologia*. Bullein, *Dialogue against the Fever Pestilence*. Scotch *Psalter*. | Acontius, *Stratagemata Satanae*. Wierus, *De Praestigiis Daemonum*. |
| Anon., *King Darius* (ent.) *Jests of Skoggan* (ent. 65/66.) Golding, tr. of Ovid, *Metamorphoses* i–iv. Googe, Palingenius, *Zodiac* i–xii. Grafton, *Manual of the Chronicles*. Jewel, *Reply unto Mr. Harding's Answer*. Thos. Peend, *Hermaphroditus and Salamacis* (after Ovid). Stow, *Summary of English Chronicles*. | Caro, *Eneide* (blank verse translation). Cooper, *Thesaurus Linguae Romanae et Britannicae*. More, *Opera*. Ronsard, *Abrégé de l'Art Poëtique*. Telesius, *De Rerum Natura*. |
| Wm. Adlington, tr. of Apuleius, *Golden Ass*. Drant, tr. from Horace, *A Medicinable Moral*. Lindsay, *Deploration of Q. Magdalene*. Wm. Painter, *Palace of Pleasure*. [Clement Robinson?] *Very Pleasant Sonnets* (ent. perhaps = *Handful of Pleasant Delights*, 1584). Jn. Studley, tr. Seneca, *Agamemnon* and *Medea*. | Bodinus, *Methodus ad facilem Historiarum Cognitionem*. Alanus Copus (= Nicholas Harpsfield) *Sex Dialogi*. Luther, *Tischreden*. |
| Wm. Allen, *Treatise in Defence of the Power of the Priesthood*. Anon., *Merry Tales Newly Imprinted . . . by Skelton*. Drant, tr. of Horace, *Art of Poetry, Epistles, and Satires*. Geoff. Fenton, *Certain Tragical Discourses*. Golding, tr., *Metamorphoses* i–xv. Thos. Harman, *A Caveat for Common Cursetors*. Jewel, *Defence of the Apology*. Mulcaster, tr. of Fortescue's *De Legibus* (w. 1460?). Painter, *Second Tome of the Palace of Pleasure*. Parker, tr. of Ælfric, *A Testimony of Antiquity*. Thos. Paynell, tr. of *Amadis*. Turberville, *Epitaphs, Epigrams; Songs and Sonnets*; tr. *Heroical Epistles of Ovid*; tr. *Eclogues* of Mantuan. Wedderburns, *Gude and Godlie Ballatis*. | Lat. tr. of *Eulenspiegel*. Pomponatius, *De Admirandorum Effectuum Causis*. |
| Bannatyne MS. (w.)? 'Bishops' Bible.' Thos. Howell, *Arbour of Amity*; *New Sonnets and Pretty Pamphlets*. Lindsay, *Works*. North, *Dial of Princes with Amplification of a IVth Book*. Skelton, *Pithy Works of*. Edm. Tilney, *Discourse of Duties in Marriage*. Turberville, tr. of Mancinus, *Plain Path to Perfect Virtue*. | Thos. Smith, *De Recta et Emendata Linguae Anglicae Scriptione*. Garnier, *Porcie*. |

| Date | Public Events | Private Events |
|---|---|---|
| 1569 | Unsuccessful rebellion of Norfolk and the Northern Earls. | Barn. Barnes b.?<br>Jn. Davies b.<br>Góngora b.<br>Marini b.<br>Last performance of York Corpus Christi plays. |
| 1570 | The Pope excommunicates and deposes Elizabeth. | Thos. Middleton b.?<br>Sam. Rowlands b.? |
| 1571 | Battle of Lepanto. Ridolfi Plot. Diplomatic relations with Spain broken off. XXXIX Articles enforced. It becomes treasonable to introduce into England Bulls or Instruments of Reconciliation with Rome. Walter Strickland suspended from H. of Commons for introducing a bill for reform of the Prayer Book. | Jewel d.<br>Benvenuto Cellini d.<br>Keppler b. |
| 1572 | Gregory XIII succ. Massacre of St. Bartholomew. Treaty of Blois. Elizabeth forbids introduction in the Commons of Bills on religion unless previously approved by the Clergy. | Knox d.<br>Ramus murdered.<br>Donne b.<br>B. Jonson b.<br>Unpatronized troops of actors declared rogues and vagabonds. |
| 1573 | Siege of La Rochelle. Relations with Spain resumed. | Wm. Laud b. |
| 1574 | Anglo-Spanish Treaty of Bristol. Persecution of Papists in England. | Richd. Barnfield b.<br>Joseph Hall b. |

| *English and Scotch Texts* | *Greek, Latin, and Continental Vernaculars* |
|---|---|
| Grafton, *Chronicle at Large*. Hawkins, *True Declaration of the Troublesome Voyage*. Thos. Newton, tr. Cicero, *Paradoxa Stoicorum* and *De Senectute*. Van der Noodt, *A Treatise &c*. Painter, 2nd edn. of *Palace of Pleasure*, vol. i. J. Sandford, tr. of Agrippa, *de Vanitate*. Thos. Underdowne, tr. Heliodorus, *Ethiopian History*; tr. Ovid, *Ibis*. Preston, *Cambyses* (w.)? | Mercator, *Chronologia*.<br>Baptista Porta, *Magia Naturalis*. |
| Ascham, *The Schoolmaster*. Foxe, *Ecclesiastical History*. Googe, tr. of Naogeorgus, *The Popish Kingdom*. Jn. Lesley, *History of Scotland* (w.). *North*, tr., *Moral Philosophy of Doni*. Tusser, *Hundred good Points . . . married unto a Hundred good Points of Housewifery*. Wilson, tr., *Three Orations of Demosthenes*. | Castelvetro, *Poetica d'Aristotele*.<br>Ortelius, *Theatrum Orbis Terrarum*. |
| Jn. Bridges, *Sermon at Paul's Cross*. Latimer, *Fruitful Sermons*. | |
| Churchyard, tr., Ovid, *Tristia* i–iii. Fenton, *Monophylo*. Jn. Field and Thos. Wilcox, *Admonition to Parliament*. Gilbert, *Q. Elizabeth's Achademy* (w.). Latimer, *Seven Sermons before the Duchess of Suffolk*. Pseudo-Cartwright, *Second Admonition*. R. H., tr. L. Lavater, *Of Ghosts and Spirits*. Tyndale, Frith, and Barnes, *Whole Works*. Whitgift, *Answer to a Libel entitled an Admonition*. Wilson, *Discourse upon Usury*. | PLUTARCH II.<br>Camoens, *Lusiadas*.<br>Matt. Parker, *De Antiquitate Britannicae Ecclesiae &c*.<br>Ronsard, *Franciade* i–iv.<br>Jean de la Taille, *Saül le Furieux*. |
| Jn. Bridges, *Supremacy of Christian Princes*. Cartwright, *Reply to an Answer*. Gascoigne, *Hundred Sundry Flowers*. Tusser, *Five Hundred Points*. | Bandello, *Novelle* iv.<br>Desportes, *Diane, &c*.<br>Du Bartas, *Judith*.<br>François Hotman, *Franco-Gallica*.<br>Tasso, *Aminta*. |
| Cartwright, tr. of Travers, *Explicatio* as *A Full and Plain Declaration*. Jn. Higgins, *The First Part of the Mirror for Magistrates*. Parker, tr. of Asser's *Life of Alfred*. Barn. Rich, *Dialogue between Mercury and an English Soldier* ? Reg. Scott, *Platform of a Hop-Garden*. Whitgift, *Defence of the Answer*. | Walter Travers, *Ecclesiasticae Disciplinae Explicatio*.<br>Garnier, *Cornelie*. |

| *Date* | *Public Events* | *Private Events* |
|---|---|---|
| 1575 | New Poor Law. Anabaptists burnt in England. | M. Parker d.<br>Thos. Heywood b.?<br>Jn. Marston b.?<br>Sam. Purchas b.?<br>Cyril Tourneur b.<br>'Princely Pleasures' at Kenilworth Castle. |
| 1576 | Sack of Antwerp. Grindal Archb. of Canterbury. Frobisher's voyage begins. Priests from Douai arrive in England. | Bullein d.<br>Cardan d.<br>Eden d.<br>Titian d.<br>'The Theatre' built in London. |
| 1577 | Drake's voyage round the world begins. | Gascoigne d.<br>Thos. Smith d.<br>Robt. Burton b.<br>Van Helmont b.<br>Curtain Theatre opened.<br>Blackfriars Theatre opened. |
| 1578 | The Pope sends Thos. Stukeley to Ireland. | Drant d.?<br>Roper d.<br>Wm. Harvey b.<br>Geo. Sandys b.<br>Jn. Heywood d. |
| 1579 | Organization of a Jesuit mission to England. Simier visits Elizabeth to treat of a m. between her and Alençon, Duke of Anjou. | Jn. Fletcher b.<br>*Hymenaeus* (a.).<br>Legge, *Richardus Tertius* (a.). |

| *English and Scotch Texts* | *Greek, Latin, and Continental Vernaculars* |
| --- | --- |
| Breton, *Small Handful of Fragrant Flowers.* Cartwright, *Second Reply.* Churchyard, *First Part of Chips.* Gascoigne, *Posies.* Golding, tr. of Beza, *Abraham's Sacrifice.* Fenton, *Golden Epistles.* Higgins, *Mirror for Magistrates* (= 1574, with an addition). Robt. Laneham, *A Letter.* Painter, *Palace of Pleasure* (2 vols.). Rolland, *Court of Venus.* Gascoigne, *Noble Art of Venery, Book of Falconry.* Mr. S, *Gammer Gurton's Needle.* | Ronsard, *Sonnets pour Hélène.*<br>Tasso, *Gerusalemme Liberata.* |
| Gascoigne, *Delicate Diet for Dainty-Mouthed Drunkards*; *Drum of Doomsday*; *Princely Pleasures of . . . Kenilworth*; *Spoil of Antwerp*; *Steel Glass* and *Philomene.* Gilbert, *Discourse of a New Passage. Paradise of Dainty Devices.* Robt. Peterson, tr. of Della Casa, *Galateo.* Pettie, *The Petite Palace of Pettie his Pleasure.* Wapull, *The Tide Tarrieth No Man.* Whetstone, *Rock of Regard.* | XENOPHON.<br>Lat. tr. of Xenophon.<br>Bodin, *La République.*<br>'Junius Brutus', *Vindiciae contra Tyrannos.*<br>L. Pasqualigo, *Il Fedele.* |
| Cartwright, *The Rest of the Second Reply.* Breton, *Flourish upon Fancy, Works of a Young Wit.* Googe, tr. of Heresbachius, *Four Books of Husbandry.* Gosson, *Treatise wherein Dicing &c. . . . are reproved* (ent.). Jn. Grange, *Golden Aphroditis.* Holinshed, the *first* and *last* volumes of the *Chronicle.* Tim. Kendall, *Flowers of Epigrams.* Hen. Peacham, *Garden of Eloquence.* Richd. Willes and Eden, *History of Travel.* | Gab. Harvey, *Ciceronianus.*<br>Lat. tr. of Psellus, *De Operatione Daemonum.* |
| Geo. Beste, *True Discourse* (of Frobisher's Voyage). Churchyard, *Lament . . . of Woeful Wars in Flanders. Gorgeous Gallery of Gallant Inventions.* Lyly, *Euphues, the Anatomy of Wit. Mirror for Magistrates* (Thos. Blennerhasset's *Second Part*). Barn. Rich, *Alarm to England.* Jn. Rolland, *The Seven Sages.* | Du Bartas, *Sepmaine.*<br>Mercator, *Tabulae Geographicae.* |
| Churchyard, *Misery of Flanders.* Fenton, tr. of Guicciardini's *History.* Gosson, *School of Abuse*; *Ephemerides of Phialo* and *Apology for the School.* Lodge [*Honest Excuses*]. Munday, *Mirror of Mutability.* North, tr. of Plutarch, *Lives of the Noble Grecians and Romans.* Pitscottie, *History* (w. not later than this). Spenser, *Shepherds' Calendar.* | Jn. Lesley, Lat. tr. of his *History* (1570), *De Origine, Moribus et Rebus Scotorum.* |

| *Date* | *Public Events* | *Private Events* |
|---|---|---|
| 1580 | Letter of Cardinal of Como to Sega pronouncing the assassination of Elizabeth lawful and meritorious. Campion and Parsons in England | Breughel d.?<br>Camoens d.<br>Holinshed d.?<br>Heinsius b.<br>Jn. Webster b.?<br>Last performance of Coventry Corpus Christi plays?<br>Teatro Olympico at Vicenza begun. |
| 1581 | Alençon in England. Secret Recusant Press set up in Essex (its products bearing the imprint 'Douai'). Execution of Edmund Campion. | Wilson d.<br>Thos. Overbury b.<br>*Pedantius* (a.)?<br>Beaujoyeulx, *Circe: Ballet Comique de la Reine* (a.). |
| 1582 | Attempted assassination of William of Orange. University of Edinburgh founded. Plague in London. Parsons expresses the opinion that nearly all Recusants desire a Spanish invasion. | Buchanan d.<br>Phineas Fletcher b.<br>Gager, *Ulysses Redux* (a.).<br>Seb. Westcott d. |
| 1583 | Whitgift Archb. of Canterbury. Irish rebellion defeated. Two plots, that of the Recusant Jn. Somerville and that of Francis Throckmorton. Court of High Commission organized. | Gilbert d.<br>Massinger b.<br>Grotius b.<br>Queen's Players formed.<br>Gager, *Rivales* and *Dido* (a.). |
| 1584 | William of Orange assassinated. Mendoza expelled from England. 'Bond of Association' formed to protect the Queen. Ralegh's failure in Virginia. | Alex. Scott d.?<br>Francis Beaumont b.<br>Jn. Selden b.<br>Lyly's *Campaspe* and *Sapho and Phao* (a.)?<br>Peele's *Arraignment of Paris* (a.).? |

| *English and Scotch Texts* | *Greek, Latin, and Continental Vernaculars* |
|---|---|
| Anon., *Manifest Detection of Diceplay*? Churchyard, *Charge* (or *Light Bundle*). Humph. Gifford, *Posy of Gilliflowers*. Harvey, *Three Proper . . . Letters*; *Two other . . . Letters*. Lyly, *Euphues and his England*. Maitland Folio (w.). Munday, *Zelauto*. Austen Saker, *Narbonus*. Stow, *Chronicles*. | Belleforest, *Histoires Tragiques*.<br>Bodin, *Démonomanie*.<br>Buchanan, *De jure Regis apud Scotos*.<br>Montaigne, *Essais* i–ii.<br>Guarini, *Il Pastor Fido* (begun). |
| Allen, *Apology of the English Seminaries*. Arthur Hall, tr. of *Iliad* i–x. Howell, *Howell his Devices*. Mulcaster, *Positions*. Pettie, tr. of Guazzo, *Civil Conversation*. Rich, *Farewell to Military Profession*, *Strange and Wonderful Adventures of Don Simonides*. Newton, etc., tr. *Seneca his Ten Tragedies*. Woodes, *The Conflict of Conscience*. | Tasso, *Gerusalemme Liberata*, revised edn.<br>Thos. Watson, Lat. tr. of Sophocles, *Antigone*. |
| Allen, *Brief Hist. of Martyrdom of XII Priests*. Breton, *Toys of an Idle Head*. Robt. Browne, *A Book which sheweth* and *Treatise of Reformation without Tarrying*. Gosson, *Plays Confuted*. Hakluyt, *Diverse Voyages*. Mulcaster, *First Part of the Elementary*. Munday, *Discovery of . . . Campion*; *English Roman Life*. Rheims N.T. Rich, *True Report of a Practice of a Papist*. R. Robinson, tr. of Leland's *Assertio*. Stanyhurst, tr. of Virgil *Aen.* i–iv. (Leyden.) Watson, *ΕΚΑΤΟΜΠΑΘΙΑ*. Whetstone, *Heptameron of Civil Discourses* (= *Aurelia*, 1593). | |
| Gilbert, *True Report of the Newfound Lands*. Greene, *Mamillia*. Wm. Hunnis, *Seven Sobs*; *Handful of Honeysuckles*; *Poor Widow's Mite*; *Comfortable Dialogues*. Jewel, *View of a Seditious Bull*; *Certain Sermons*; *Exposition on Thessalonians*. Brian Melbancke, *Philotimus*. T. Smith, *De Republica Anglorum*. Stanyhurst, *Aeneid* i–iv (London). Philip Stubbs, *Anatomy of Abuses*; *Second Part of the Anatomy*; *Rosary of Christian Prayers*. | Du Bartas, *Seconde Sepmaine*.<br>Garnier, *Tragedies*. |
| Allen, *True, Sincere, and Modest Defence* (Rouen). Wm. Fulke (or Dudley Fenner?), *Brief and Plain Declaration* (= *The Learned Discourse*). Greene, *Mirror of Modesty*; *Card of Fancy* (= *Gwydonius* or on titlepage, *Guydemus*). Hakluyt, *Western Discoveries* (w.). James VI, *Essays of a Prentice*. Lodge, *Alarm against Usurers*, | Jean Benedicti, *Somme des Pechez*.<br>Giordano Bruno, *Cena de le Ceneri*; *De la causa*; *De l'infinito*. |

| Date | Public Events | Private Events |
|---|---|---|
| 1585 | Sixtus V succ. Elizabeth refuses the sovereignty of the Netherlands. Expedition to the Netherlands under Leicester. Drake at Vigo, &c. Execution of Dr. Parry. Bachelors and undergraduates at Oxford ordered to confine themselves (in Logic) to Aristotle 'and those that defend him'. | Ronsard d.<br>Wm. Drummond b.<br>Giles Fletcher (junior) b.<br>Hooker, *On Justification* (prc.). |

| *English and Scotch Texts* | *Greek, Latin, and Continental Vernaculars* |
|---|---|
| and *Forbonius and Prisceria*. Munday, *Watchword to England*. Rich, *Second Tome of Don Simonides*. Scott, *Discovery of Witchcraft*. Jn. Southern, *Pandora*. Warner, *Pan his Syrinx*. Whetstone, *Mirror for Magistrates of Cities* and *Touchstone for the Time*. R. Wilson, *Three Ladies of London*. | |
| Greene, *Planetomachia*. Munday, tr. of L. Pasqualigo, *Fedele and Fortunio*. Peele, *Pageant for Dixi*. Stubbs, *Intended Treason of Dr. Parry*. Whetstone, *Enemy to Unthriftiness* (= *Mirror for Magistrates of Cities*, 2nd edn.); *Mirror of True Honour*. | Bruno, *Eroici Furori*.<br>Watson, *Amyntas*. |

# BIBLIOGRAPHY

CONTENTS:



ABBREVIATIONS:

| | |
|---|---|
| * | Facsimile Text |
| Adams | *Chief Pre-Shakespearean Dramas*, ed. J. Q. Adams (1924) |
| *Archiv* | *Archiv für das Studium der neueren Sprachen* |
| Bevington | D. M. Bevington, *From 'Mankind' to Marlowe* (1962) |
| Brandl | A. Brandl, *Quellen des weltlichen Dramas* (Strassburg, 1898) |
| *CBEL* | *Cambridge Bibliography of English Literature* |
| Chambers, *ES* | E. K. Chambers, *The Elizabethan Stage*, 4 vols. (1923) |
| Craik | T. W. Craik, *The Tudor Interlude* (1962) |
| *DNB* | *Dictionary of National Biography* |
| Dodsley | Dodsley's *Old Plays* (1744) |
| EETS ES | Early English Text Society (extra series) |
| EETS OS | Early English Text Society (ordinary series) |
| *ELN* | *English Language Notes* |
| *E. Studien* | *Englische Studien* |
| *E. Studies* | *English Studies* |
| Farmer | *Lost Tudor Plays*; *Tudor Facsimile Texts**; *Anonymous Plays* (four series), ed. J. S. Farmer |
| Gayley | *Representative English Comedies*, ed. C. M. Gayley, 3 vols., 1903–14 |
| H-D | W. C. Hazlitt's edn. of Dodsley's *Old Plays* (1874) |
| *HLQ* | *Huntington Library Quarterly* |

| | |
|---|---|
| *JEGP* | *Journal of English and Germanic Philology* |
| Manly | *Specimens of the Pre-Shaksperean Drama*, ed. J. M. Manly, 2 vols (1897–8) |
| Materialien | Materialien zur Kunde des älteren englischen Dramas (*continued as* Materials for the study of the old English drama) |
| *MLN* | *Modern Language Notes* |
| *MLQ* | *Modern Language Quarterly* |
| *MLR* | *Modern Language Review* |
| *MP* | *Modern Philology* |
| MSC | Malone Society Collections |
| MSR | Malone Society Reprint |
| n.d. | no date |
| *NQ* | *Notes and Queries* |
| OHEL | Oxford History of English Literature |
| *PMLA* | *Publications of the Modern Language Association of America* |
| Pollard | *English Miracle Plays*, ed. A. W. Pollard (1927) |
| *PQ* | *Philological Quarterly* |
| *RES* | *Review of English Studies* |
| S.R. | Stationers' Register |
| *Sh. Jb.* | *Jahrbuch der deutschen Shakespeare Gesellschaft* |
| *Sh. Q.* | *Shakespeare Quarterly* |
| *Sh. S.* | *Shakespeare Survey* |
| Spivack | B. Spivack, *Shakespeare and the Allegory of Evil* (1958) |
| *SP* | *Studies in Philology* |
| *STC* | Pollard and Redgrave, *Short Title Catalogue of English Books, 1475–1640* (1926) |
| STS | Scottish Text Society |
| *TLS* | *Times Literary Supplement* |

Place of publication is ordinarily given only for books published in countries other than Great Britain and the United States.

## I. GENERAL BIBLIOGRAPHY AND LITERARY HISTORY

General bibliographical aids are to be found in the Catalogues of Printed Books in major libraries, the British Museum, the Library of Congress, the Bibliothèque Nationale, in *CBEL* ed. F. W. Bateson (4 vols., 1940) with supplement, ed. G. Watson (1957), in *STC* ed. Pollard and Redgrave (1926). The indispensable work of bibliographical description is W. W. Greg, *A Bibliography of English Printed Drama to the Restoration* (4 vols., 1962). Greg's earlier *A List of English Plays Written Before 1643 and Printed Before 1700* (1900)—with its supplement: *A List of Masques, Pageants etc. Supplementary to a List of English plays* (1902)—is easier to use, but is largely superseded by A. Harbage, *Annals of English Drama* (1940), revised by S. Schoenbaum (1962; supplement, 1966). C. J. Stratman, *A Bibliography of Medieval Drama* (1954) covers the period in a mechanical and often inaccurate way. R. J. Schoeck gives an account of scholarly needs in this field in 'Research in early Tudor drama', *Opportunities for Research in Renaissance Drama* (MLA, 1956). E. K. Chamber's *The Mediaeval Stage* (2 vols., 1903), and *The Elizabethan Stage* (4 vols., 1923) collect and reduce to useful order the masses of fragmentary information that have survived. W. C. Hazlitt, *The English Drama and the Stage 1543–1664* (Roxburghe Library, 1869) is a similar compilation.

Among histories of the drama and literature of the period T. Warton's *History of English Poetry* (3 vols., 1774–81) ed. Hazlitt (4 vols., 1871) contains some things that are otherwise unknown, but is not entirely trustworthy. The compilation of J. P. Collier (*The History of English Dramatic Poetry to the time of Shakespeare and Annals of the Stage to the Restoration* (3 vols., 1831; 1879) is also, and to a much greater degree, subject to suspicion. F. G. Fleay, *A Chronicle History of the London Stage 1559–1642* (1890) and *A Biographical Chronicle of the English Drama 1559–1642* (2 vols., 1891) is also liable to scrutiny for its too great facility in solving authorship problems. The standard histories of Elizabethan Drama, e.g. A. W. Ward's (3 vols., 1875–99), F. E. Schelling's (2 vols., 1908), cover the plays in this volume fairly perfunctorily. The *Cambridge History of English Literature* (1907–16) has some valuably detailed chapters—F. S. Boas on 'Early English comedy' (vol. v), W. Creizenach on 'Miracle plays and moralities' (vol. v), J. W. Cunliffe on 'Early English tragedy' (vol. v), F. S. Boas on 'University plays' (vol. vi). Similarly valuable is C. M. Gayley's 'An historical view of the beginnings of English comedy' in his *Representative English Comedies*, vol. i (1903).

Creizenach's great *Geschichte des neueren dramas* (5 vols., Halle, 1893–1916) is still the only effective survey of European Renaissance drama: only part of vol. iv is translated (*English Drama in the Age of Shakespeare* (1916)). C. W. Wallace, *The Evolution of English Drama up to Shakespeare* (1912), is slightly tendentious. Greater concentration on the period of this volume is provided by C. F. T. Brooke, *The Tudor Drama* (1912), A. W. Reed, *Early Tudor Drama* (1926), and F. S. Boas's primer, *An Introduction to Tudor Drama* (1933). More impressionistic criticism is in J. A. Symonds, *Shakspere's Predecessors in the English Drama* (1884; 1900; 1906), J. J. Jusserand, *Le théâtre en Angleterre . . . jusq'aux prédécesseurs immédiats de Shakespeare* (Paris, 1878), and A. P. Rossiter, *English Drama from early times to the Elizabethans* (1950). E. N. S. Thompson, *The English Moral plays* (*Transactions of the Connecticut Academy of Arts and Sciences*, 1910), provides a painstaking survey of the movement from *The Castle of Perseverance* to *Liberality and Prodigality*. W. Roy Mackenzie writes on *The English Moralities from the point of view of Allegory* (1914), and Willard Thorp on *The Triumph of Realism in Elizabethan Drama 1558–1612* (1928), L. B. Campbell on *Divine Poetry and Drama in Sixteenth Century England* (1953), A. Lombardi on *Il dramma pre-Shakespeareano* (Venice, 1957). T. W. Craik's *The Tudor Interlude* (1958) is a general survey centred on stage presentation, and D. M. Bevington, *From 'Mankind' to Marlowe* (1962) deals with the theatrical structure of 'popular' plays in this period. L. B. Wright deals with 'Social aspects of some belated moralities' in *Anglia*, 1930, and elsewhere with 'The Scriptures and the Elizabethan Stage', *SP*, 1923.

## II. COLLECTIONS OF PLAYS, ANTHOLOGIES, ETC.

R. Dodsley's *Old Plays* (12 vols., 1744) was edited and enlarged by I. Reed (1780) and J. P. Collier (1825–7) and given definitive form by W. C. Hazlitt (15 vols., 1874–6). The *Materialien zur Kunde des älteren englischen Dramas* was supervised by W. Bang (Louvain, 1902–14) and its continuation, the *Materials for the Study of Old English Drama* by H. de Vocht (Louvain, 1927 et seq.). The Malone Society has been supervised successively by W. W. Greg, F. P. Wilson, and Arthur Brown (1907 et seq.). J. S. Farmer put forth *Tudor Facsimile Texts* (143 vols., 1907–14), *'Lost' Tudor Plays* (1907), *Anonymous Plays* (four series, 1905). C. M. Gayley was general editor of *Representative English*

*Comedies* (3 vols., 1903–14). See also J. M. Manly, *Specimens of the Pre-Shaksperean Drama* (2 vols., 1897–8); J. Q. Adams, *Chief Pre-Shakespearean Dramas*, 1924; F. S. Boas, *Five Pre-Shakespearean Comedies*, 1934; W. A. Armstrong, *Elizabethan History Plays* (1965); R. W. Bond, *Early Plays from the Italian*, 1911; J. W. Cunliffe, *Early English Classical Tragedies* (1912); Alois Brandl, *Quellen des weltlichen Dramas in England vor Shakespeáre* (Strassburg, 1898). Many masques, mummings, and ceremonies are preserved in J. Nichols, *Progresses of Queen Elizabeth* (3 vols., 1788–1821).

## III. PARTICULAR STUDIES

*Conventional Features*

On stylistic aspects of the drama see B. J. Whiting, *Proverbs in the Earlier English Drama* (1938); F. I. Carpenter, *Metaphor and Simile in the Minor Elizabethan Drama* (1895). The standard work on prosody is J. E. Bernard, *The Prosody of the Tudor Interlude* (1939); see also F. G. Hubbard, 'Repetition and parallelism in earlier Elizabethan drama', *PMLA*, 1905; and 'A type of blank verse line . . .', *PMLA*, 1917; E. Eckhardt, 'Die metrische Unterscheidung von Ernst und Komik in den englischen Moralitäten', *E. Studien*, 1927; J. F. MacDonald, 'The use of prose in English drama before Shakespeare', *UTQ*, 1932/33.

Turning to structural features—debate-structure is discussed by J. H. Hanford, 'The debate element in Elizabethan drama', *Anniversary Papers for G. L. Kittredge* (1913). H. Walther treats the debate more widely in *Das Streitgedicht* (Munich, 1920). See also E. Waith, 'Controversia in the English drama . . .', *PMLA*, 1953. On dumb-show, see B. R. Pearn, 'Dumb-show in Elizabethan drama', *RES*, 1935, and D. Mehl, *The Elizabethan Dumb Show* (1965).

On character-types—the 'Vice' receives his standard treatment in L. Cushman, *The Devil and the Vice in the English Dramatic Literature before Shakespeare* (Halle, 1900); more modern and more critical is B. Spivack, *Shakespeare and the Allegory of Evil* (1958). See also T. E. Allison, 'The paternoster play and the origin of the Vices', *PMLA*, 1924; R. Withington, 'The Vice: ancestry, development', in *Excursions in English Drama* (1937); F. H. Mares, 'The origin of the figure called "The Vice" in Tudor Drama', *HLQ*, 1958; R. Weimann, 'Redekonventionen des Vice . . .', *Zeitschrift für Anglistik und Amerikanistik*, 1967.

On the braggart see H. Graf, *Der Miles Gloriosus im englischen Drama* (1902), and D. C. Boughner, *The Braggart in Renaissance Comedy* (1954). On fools, see O. M. Busby, *Studies in the Development of the Fool in Elizabethan Drama* (1923), B. Swain's *Fools and Folly during the Middle Ages and the Renaissance* (1932), E. Welsford's *The Fool* (1935). See also R. H. Goldsmith, *Wise Fools in Shakespeare* (1955), and W. Kaiser, *Praisers of Folly* (1963). J. R. Moore discusses 'Ancestors of Autolycus in the English Moralities' in the *Heller Memorial Volume* (Washington University Studies IX, 1922). E. P. Vandiver deals with 'The Elizabethan dramatic parasite' in *SP*, 1935. E. Eckhardt describes *Die lustige Person im älteren englischen Drama* (1902). F. W. Moorman writes on 'The pre-Shakespearean Ghost', *MLR*, 1906, W. A. Armstrong on 'The Elizabethan conception of the tyrant', *RES*, 1946.

On connections between Morality plays and Elizabethan drama see H. Craig, *Sh. Q.*, 1950; I. Ribner, 'Morality roots of the Tudor history play', *Tulane Studies in English*, 1954; H. M. V. Matthews, *Character and Symbol in Shakespeare's Plays* (1962), R. G. Hunter, *Shakespeare and the Comedy of Forgiveness* (1965), and several essays in *Elizabethan Theatre*, ed. Brown and Harris (1966).

*Court Drama*

Court disguisings and masques are frequently described in Hall's Chronicle (1548; rept. 1809 [ed. H. Ellis]). A. Feuillerat has printed *Documents relating to the Revels at Court in the time of King Edward VI and Queen Mary* (Materialien, vol. 44, 1914) and *Documents relating to the office of the Revels in the time of Queen Elizabeth* (Materialien, vol. 21, 1908). Further information about these occasions may be found in *Letters and Papers of Henry VIII* (ed. J. Gairdner), and in the *Calendar of State Papers* (Domestic). On the masque proper, see R. Brotanek, *Die Englischen Maskenspiele* (1902); P. Reyher, *Les Masques Anglais* (1909); E. Welsford, *The Court Masque* (1927). See also Mary S. Steele, *Plays and Masques at Court* (1926); and L. M. Ellison, *The Early Romantic Drama at the English Court* (1917). C. R. Baskervill writes on 'Some evidence for early romantic plays in England', *MP*, 1916/17 (229–51; 467–512). See also W. A. Neilson, *Origins and Sources of the Courts of Love* (1899), T. F. Crane, *Italian Social Customs of the Sixteenth Century* (1920), S. Anglo, 'The Court festivals of Henry VII', *Bulletin of the John Rylands Library*,

1960, Frances Yates, 'Elizabethan chivalry: the romance of the accession day tilts', *Journal of Warburg and Courtauld Institutes*, 1957, J. Jacquot (ed.) *Fêtes de la Renaissance* (2 vols., Paris, 1956).

*Academic Drama*

F. S. Boas's *University Drama in the Tudor Age* (1914) remains the standard work in this field. T. H. V. Motter's *The School Drama in England* (1929) has some merits as a survey. L. E. Tanner, *Westminster School* (1951) has an appendix of 'Notes and documents illustrating the early history of the play'. There is a checklist of Latin plays in England, by Leicester Bradner, in *PMLA*, 1943, revised in *Studies in the Renaissance*, 1957. See also G. R. Churchill and W. Keller, 'Die lateinischen Universitäts-Dramen in der Zeit der Königin Elisabeth', *Sh. Jb.*, 1898, with summaries of the plays described. C. F. T. Brooke writes on 'Latin drama in Renaissance England' in *ELH*, 1946. See also 'Neo-Latin drama: two views of opportunities' by L. A. Schuster and L. Bradner, *Renaissance Drama*, 1963. G. C. Moore Smith lists *College Plays performed in the University of Cambridge* (1923) and in MSC ii (1923), described 'The academic drama in Cambridge'. R. E. Alton has printed *Oxford Dramatic Records* (from four colleges) in MSC v (1959). There is a bibliography of drama and the Inns of Court by D. S. Bland, published in *Research Opportunities in Renaissance Drama*, ix (1966). See also A. W. Green, *The Inns of Court and Early English Drama* (1931) and R. J. Schoeck, 'Early Tudor drama and the Inns of Court', *American Society for Theatre Research Newsletter*, November 1957. J. W. Cunliffe, *Early English Classical Tragedies* (1912), comments on Inns of Court staging.

*Theatres and Actors*

J. T. Murray, *English Dramatic Companies 1558–1642* (2 vols., 1910) and E. K. Chambers, *Mediaeval Stage* (2 vols., 1903) and *Elizabethan Stage* (4 vols., 1923) have been brought up to date, in part, in G. Wickham, *Early English Stages* (1959, 1963 et seq.; in progress). Additional light is thrown on the stage tradition in G. Kernodle, *From Art to Theater* (1944) in R. Withington, *English Pageantry* (2 vols., 1918–20), in R. Southern, 'The contribution of the Interludes to Elizabethan staging', *Essays . . . in Honour of Hardin Craig* (1962), in lectures by Southern and S. Anglo, printed in *Le Lieu Théâtral à la Renaissance* (ed. Jacquot) (Paris, 1964), and in M. D. Anderson, *Drama and Imagery in Eng-*

*lish Medieval Churches* (1964). See also George and Portia Kernodle, 'Dramatic elements in the medieval tournament', *Speech Monographs* ix (1942), and R. S. Loomis, 'The allegorical siege in the art of the Middle Ages', *American Journal of Archaeology*, Second Series, xxiii (1919); E. Lauf, *Die Bühnenanweisungen in den englischen Moralitäten und Interludien bis 1570* (Münster, 1932). New records have been printed in D. J. Gordon and Jean Robertson, *A Calendar of Dramatic Records in the Books of the Livery Companies of London 1485–1640* (MSC iii, 1954; v, 1959), in *The Dramatic Records of the City of London*, MSC i (1), 1908 and MSC ii (3), 1931. Records relating to the first Blackfriars stage are printed in MSC ii (1), 1913. See also C. T. Prouty, 'An early Elizabethan playhouse', *Sh. S.*, 1953. H. N. Hillebrand, *The Child Actors* (1926), remains the standard monograph on this aspect of the theatrical tradition. The essays of W. J. Lawrence (*Pre-Restoration Stage Studies* (1927), *Those nutcracking Elizabethans* (1935), *The Elizabethan Playhouse and Other Studies* (2 vols., 1912–13)) remain usefully informative. On actors see A. F. Leach, 'Some English plays and players, 1220–1548' in *An English Miscellany presented to Dr. F. J. Furnivall* (1901), E. Nungezer, *A Dictionary of Actors*, (1929), M. C. Bradbrook, *The Rise of the Common Player* (1962).

On court staging, etc., see A. A. Helmholtz-Phelan, 'The staging of court drama to 1595', *PMLA*, 1909; T. S. Graves, *The Court and the London Theatres during the Reign of Elizabeth* (1913); J. H. McDowell, 'Tudor court staging: a study in perspective', *JEGP*, 1945; M. Paterson, 'The stagecraft of the Revels Office during the reign of Elizabeth', *Studies in the Elizabethan Theatre*, ed. Prouty (1961), and L. B. Campbell, *Scenes and Machines on the English Stage during the Renaissance* (1923). The earlier studies should be read with the discussion of Wickham in mind.

On control of the stage see E. K. Chambers, *Notes on the History of the Revels Office under the Tudors* (1906); V. C. Gildersleeve, *Government Regulation of the Elizabethan Drama* (1908); A. Feuillerat, *Le Bureau des menus-plaisirs et la mise-en-scène à la cour d'Élizabeth* (Louvain, 1910).

*Stage-Music*

John Stevens, *Music and Poetry in the Early Tudor Court* (1961), is a useful up-to-date account of the society that produced much of the surviving music in this period. The fifth edition of *Grove's Dictionary of Music and Musicians* (ed. Blom, 10 vols., 1954–61) has many recently revised and relevant articles. See also J. Stevens, 'Music in Medieval drama', *Proceedings of Royal Musical Association*, 1958; Grattan Flood, *Early Tudor Composers* (1925); J. Pulver, *A Biographical Dictionary of Old English Music* (1927); [G. E. P. Arkwright], 'Early Elizabethan stage music', *Musical Antiquary* (1909/10 and 1912/13).

*Literary Forms*

On general generic theory see M. Doran, *Endeavors of Art* (1954). On tragedy see R. Fischer, *Zur Kunstentwicklung der englischen Tragödie von ihren ersten Anfangen bis zu Shakespeare* (Strassburg, 1893). H. H. Adams, *English Domestic or Homiletic Tragedy, 1575–1642* (1943); Willard Farnham, *The Medieval Heritage of Elizabethan Tragedy* (1936); W. Peery, 'Tragic retribution in the 1559 *Mirror for Magistrates*', *SP*, 1949; W. Clemen, *English Tragedy before Shakespeare* (1955; 1961); P. Edwards, *Thomas Kyd and Early Elizabethan Tragedy* (1966).

On comic form see O. E. Winslow, *Low Comedy as a Structural Element in English Drama* (1926); T. W. Baldwin, *William Shakspere's Five Act Structure* (1947); M. T. Herrick, *Comic Theory in the Sixteenth Century* (1950). On tragicomedy see E. Ristine, *English Tragicomedy* (1910), and M. T. Herrick, *Tragicomedy: its origin and development in Italy, France, and England* (1955). On satire see E. M. Campbell, *Satire in the Early English Drama* (1914), supplemented by J. Peter, *Complaint and Satire* (1956).

*Foreign Influence*

Translations from the Classics are listed in H. R. Palmer, *List of English Editions and Translations of Greek and Latin Classics before 1641* (1911), and H. B. Lathrop, *Translations from the Classics into English* (1933); M. A. Scott lists *Elizabethan Translations from the Italian* (1916). The influence of Terence is writ large in T. W. Baldwin's *William Shakspere's Small Latine and Less Greke* (2 vols., 1944) and *Five Act Structure* (1947). C. H. Herford's *Studies in the Literary Relations of England and Germany in the Sixteenth Century* (1886) discusses the Reformation and the pedagogic plays of the period. On Prodigal Son plays see H. Holstein, *Das Drama vom verlorenen Sohn* (Halle, 1880) and F. Spengler, *Der verlorene Sohn im Drama des XVI Jahrhunderts* (Innsbruck, 1888). See also A. B. Feldman, 'Dutch Humanism and the Tudor dramatic tradition', *NQ* 16 August 1952. The *Acolastus* of Guilhelmus Gnapheus (or Fullonius)—born Willem de Volder—(1493–1568) was first published in 1529. It is edited by J. Bolte, 1891, by P. Minderaa (with Dutch translation) (1956), and by W. E. D. Atkinson (with English translation) (1964). The 'Ekphrasis' or expanded translation by John Palsgrave (1540) is edited by P. L. Carver for EETS (o.s. 202, 1937). For Sixt Birck or Betulius (1500–54) see H. Levinger, *Augsburger Schultheater* (Berlin, 1931), and A. W. Reed, 'Sixt Birck and Henry Medwall', *RES*, 1926. The plot of *Sapientia Salomonis* (1547) is summarized in *Sh. Jb.*, 1898; the play is edited by E. R. Payne, 1938. For the performance at Westminster School in 1566 see L. Tanner, *Westminster School: a History* (1934; 1951). Ravisius Textor or Jean Tissier (Tixier) de Ravisy (1480–1524) had his *Dialogi* first printed in 1530. Two Textor dialogues used by Tudor dramatists are printed by F. Holthausen in *E. Studien*, 1902. One dialogue is englished by T. Heywood in his *Pleasant Dialogues and Drammas* (1637). G. C. Moore Smith in *Fasciculus Joanni Willis Clark Dicatus* (1909) shows that a Textor dialogue was acted at Queens' College, Cambridge, in 1543. Creizenach, *Geschichte*, II. 56–62, discusses the European diffusion of Textor. See also J. Vodoz, *Le Théâtre latin de Ravisius Textor* (1898). For Thomas Kirchmeyer or Naogeorgus (1511–63) see F. Wiener, *Naogeorgus im England der Reformationzeit* (Berlin, 1913). For the performance of his *Pammachius* (1538) at Cambridge in 1545 see F. S. Boas, *University Drama*. His *Mercator* is edited by J. Bolte (Leipzig, 1927). The biblical plays of Cornelius Schonaeus (Schoon *or* de Schoone) (1541–1611) are described by Otto Francke, *Terenz und die lateinische Schulcomoedie in Deutschland* (Weimar, 1877). There are translations into English in Bodleian Rawlinson MSS. For Georgius Macropedius (Joris van Langenfeld) (1475–1558) see C. H. Herford, *Literary Relations*, and Max Herrman and S. Szamatolski, *Lateinische Literaturdenkmäler des XV u. XVI Jahrhunderts* (Berlin, 1891).

For French influence see s.v. Heywood. See also Grace Frank, *The Medieval French Drama* (1954); J. Mortensen, *Le Théâtre français au moyen âge* (Paris, 1903); L. Petit de Julleville, *Répertoire du théâtre comique en France au moyen âge* (Paris, 1886); E. Fournier, *Le Théâtre français avant la Renaissance 1450–1550* (Paris, n.d. [1872]); Gustave Cohen, *Le Théâtre en France au moyen âge* (2 vols., Paris, 1928, 1931).

For the influence and example of foreign stages see J. W. Cunliffe, 'Italian prototypes of masque and dumb-show', *PMLA*, 1907; W. H. Shoemaker, *The Multiple Stage in Spain during the Fifteenth and Sixteenth Centuries* (1935); and Gustave Cohen, *Histoire de la mise en scène dans le théâtre religieux français du moyen âge* (Paris, 1906).

*Seneca and his influence*

Useful accounts of Seneca's plays and their aesthetic appear in L. Herrmann, *Le théâtre de Sénèque* (Paris, 1924); N. T. Pratt, *Dramatic Suspense in Seneca* (1939); in Berthe Marti, 'Seneca's tragedies: a new interpretation', *Transactions of the American Philological Association*, 1945; and 'Prototypes of Seneca's Tragedies', *Classical Philology*, 1947, Charles Garton, 'The background to character portrayal in Seneca', *Classical Philology*, 1959; D. Henry and B. Walker, 'Seneca and the *Agamemnon*: some thoughts on tragic doom', *Classical Philology*, 1963; Regenbogen, *Schmerz und Tod in den Tragödien Senecas* (Hamburg, 1930); C. W. Mendell, *Our Seneca* (1941).

The general aspect of Seneca's influence is canvassed in *Les Tragédies de Sénèque et le théâtre de la Renaissance*, ed. J. Jacquot (Paris, 1964). Creizenach, op. cit., deals with his European influence. See also P. Stachel, *Seneca und das deutsche Renaissancedrama* (1907). On his English influence the standard work is J. W. Cunliffe's *The Influence of Seneca on Elizabethan Tragedy* (1893); contradicted by H. Baker, *Induction to Tragedy* (1939) and G. K. Hunter, 'Seneca and the Elizabethans', *Shakespeare Survey*, 1967; and modified by much subsequent work, for example, J. M. Manly, 'The influence of the tragedies of Seneca upon early English drama' in F. J. Miller's translation of Seneca's tragedies (1907); A. M. Witherspoon, *The Influence of Robert Garnier on Elizabethan Drama* (1924); H. B. Charlton, *The Senecan Tradition in Renaissance Tragedy* (1946), reprinted from *The Poetical Works of Sir William Alexander* (STS, 1921); Peter Ure, 'On some differences between Senecan and Elizabethan tragedy', *Durham University Journal*, 1948; A. H. Gilbert, 'Seneca and the criticism of Elizabethan tragedy', *PQ*, 1934; G. L. Evans, 'Shakespeare, Seneca, and the kingdom of violence', in *Roman Drama*, ed. T. A. Dorey and D. R. Dudley (1965); H. W. Wells, 'Senecan influence on Elizabethan tragedy', *Shakespeare Association Bulletin*, 1944; T. Spencer, *Death and Elizabethan Tragedy* (1936); T. S.

Eliot, 'Shakespeare and the Stoicism of Seneca', *Selected Essays* (1932). Gisela Dahinten, *Die Geisterszene in der Tragödie vor Shakespeare: zur Seneca-nachfolge* . . . (Göttingen, 1958).

The Tudor translations of Seneca were by Jasper Heywood (*Troades* [1559], *Thyestes* [1560], *Hercules Furens* [1561]), John Studley (*Agamemnon* [1566], *Medea* [1566]), Thomas Nuce (*Octavia* [1566]), Alexander Nevyle (*Oedipus* [1563]). These were all reprinted (in revised forms) in Thomas Newton's collection, *Seneca his tenne tragedies* (1581), and completed by the addition of hitherto unpublished translations of *Thebais* by Newton, and of *Hippolytus* and *Hercules Oetaeus* by Studley. The three Jasper Heywood volumes have been reprinted in *Jasper Heywood and his Translation of Seneca's Troas, Thyestes and Hercules Furens*, ed. H. de Vocht (Materialien, 1913) and the two Studley volumes in *Studley's translations of Seneca's Agamemnon and Medea*, ed. E. M. Spearing (Materialien, 1913). E. M. Spearing has written on *The Elizabethan Translations of Seneca's tragedies* (1912) and E. Jockers on *Die englischen Seneca-Uebersetzer des 16 Jahrhunderts* (Strassburg, 1909). The Newton collection was reprinted by J. Legh for the Spenser Society (2 vols., 1887) and in 'The Tudor translations' (2 vols., 1927) with an introduction by T. S. Eliot ('Seneca in Elizabethan translation', reprinted in *Selected Essays*, 1932).

## IV. INDIVIDUAL AUTHORS

R.B. (*fl.* 1559–67)

The author of *Apius and Virginia*, sometimes identified with Richard Bower (Master of the Chapel children, 1561–6). The play, written *c.* 1559–67 (S.R. 1567), was published 1575. It is reprinted in H-D 4, Adams, and MSR, 1911. See also Spivack, pp. 269 ff., etc.

BALE, JOHN (1495–1563)

The most complete account of Bale's life and publications appears in W. T. Davies's article in *Oxford Bibliographical Society Proceedings*, v (1940). This is supplemented by notes by J. F. Mozley (*Notes and Queries*, 1945). C. S. Lewis (OHEL) has a review of his non-dramatic writings. J. S. Farmer has printed the *Dramatic Writings* (1907). It is clear from his own *Catalogus* that many of Bale's plays have perished.

*Kynge Johan* (first version 1538; second 1558–62)—manuscript in Huntington Library—was edited by J. P. Collier (Camden Society, 1838), in Manly I, edited Bang, Materialien*, 1909, edited J. H. P. Pafford and W. W. Greg, MSR, 1931, edited W. A. Armstrong, *Elizabethan History Plays* (1965). Extracts appear in A. W. Pollard's *English Miracle Plays* (1927). For omissions in the earlier editions see C. E. Cason in *JEGP*, 1928, corrected by J. H. P. Pafford in *JEGP*, 1931. See also H. Barke, Bales '*Kynge Johan*' . . . (Würzburg, 1937).

Bale's *Tragedye or enterlude manyfestyng the chefe promyses of God unto man* (1538) was printed in [1547?] and 1577; edited in Marriott's *Collection of English Miracle Plays* (Basel, 1838) and in Dodsley's *Old Plays* (1744 et seq.); edited Emrys Jones, Erlangen, 1909, and Ernest Rhys (*Everyman with other Interludes*, 1909), Farmer (1908)*. *A brefe Comedy or enterlude concernynge the temptacyon of our Lorde* (1538) was reprinted in Grosart's *Miscellanies of the Fuller Worthies' Library*, vol. i (1870); Farmer (1909)*; edited Paul Schwemmer, Nürnberg, 1919. *A comedy concernynge thre lawes of nature Moses, and Christ* (1538) was edited by A. Schroeer in *Anglia*, 1882, Farmer (1908)*. See also Bevington, pp. 128 ff., etc., Craik, pp. 73 ff., etc. *A Briefe comedy or Enterlude of Johan Baptystes* (1538), reprinted in *Harleian Miscellany*, vol. i (1744). A fragmentary manuscript in the Folger Library is in MSR, 1912, with the title 'The Resurrection of Our Lord', edited by J. D. Wilson and Bertram Dobell—and may be by Bale.

### BUCHANAN, GEORGE (1506–82)

There is a bibliography by D. Murray in *George Buchanan: Glasgow Quatercentenary Studies*, 1907. His translations of the *Medea* and the *Alcestis* of Euripides were printed in Paris, 1544 and 1556 respectively, and in numerous editions of the *Opera Omnia*. Of his original Latin plays, *Jephthes* was printed in Paris, 1554, etc., translated into English by W. Tait, 1750, C. C. Truro, 1853, A. G. Mitchell, n.d. [1903]; *Baptistes* was written *c*. 1540, published London, 1577, etc., tr. English anon. [sometimes supposed to be John Milton] 1642 under the title *Tyrannical-Government Anatomised* (reprinted 1740, 1907); translated also by A. Gibb (with *Jephthes*), 1870, A. G. Mitchell, 1904. *Sacred Dramas* are translated (in verse) by A. Brown (1906).

CORNISH (CORNYSHE), WILLIAM (*c.* 1468?–1523)

For biographical details see *DNB*, *Grove's Dictionary of Music* (fifth edition, ed. Blom, 1954–61), J. Pulver, *A Biographical Dictionary of Old English Music* (1927). Chambers, *ES*, ii. 29, follows Grove in what is probably an error—supposing that there were two men involved in this one career.

C. W. Wallace (*Evolution of the English Drama*, 1912) supposed Cornish to be the author of many early plays; but his views have not been accepted. No plays definitely written by him have survived, but his work as impresario, actor, choir-master, musician is recorded in Hall's Chronicle and elsewhere. His *Treatise bitwene Trouth and Enformacion* was printed in Skelton's *Pithy . . . works* (1568) and (from BM Royal MS. 18 D 11) by Flügel in *Anglia*, 1892. See also S. Anglo in *RES*, 1959.

EDWARDS, RICHARD (1524–66)

Edwards's life is described in Leicester Bradner, *The Life and Poems of Richard Edwards* (1927). Warton's *History* devotes a chapter to him. For Edwards's relationship to the miscellany volume *The Paradise of Dainty Devices* (1576) see the edition by Hyder Rollins (1927), with corrections in Rollins's article in *J. Q. Adams Memorial Studies* (1948). His music is dealt with in W. G. Grattan Flood's *Early Tudor Composers* (1925) and in *Grove's Dictionary of Music* (fifth edition, 1954–61). His position as master of the Children of the Chapel Royal (1561–6) is the subject of discussion in H. N. Hillebrand's *The Child Actors* (1926). His play *Palamon and Arcite*, performed before the Queen in Christ Church Hall, Oxford, in 1566, is lost; but descriptions of the occasion and the plot survive among the *commentarii* of J. Bereblok, reprinted in C. Plummer, *Elizabethan Oxford* (Oxford Historical Society, 1887), and in other contemporary accounts. See W. Y. Durand in *PMLA*, 1905, and (especially) F. S. Boas, *University Drama in the Tudor Age* (1914). Hyder Rollins prints in *RES*, 1928, a manuscript elegy which appears to have been sung in this play. *Damon and Pithias* was probably acted at Court, Christmas 1564; Stationers' Register 1567; published 1571; 1582. Reprinted in H-D 4, Farmer (1906, 1908*), Adams, MSR (1957).

Its source in Elyot's *Boke named the Governour* is discussed by Bradner, op. cit. W. A. Armstrong has written on '*Damon and*

*Pithias* and Renaissance theories of tragedy', *E. Studies*, 1958; J. L. Jackson has 'Three notes on *Damon and Pithias*' in *PQ*, 1950. L. J. Mills discusses the theory of friendship behind the play in 'Some aspects of *Damon and Pithias*', *Indiana University Studies* (1927). See also A. Holaday in *JEGP*, 1967. J. W. Cunliffe discusses the two plays in his preface to *Early English Classical Tragedies*.

FARRANT, RICHARD (*c.* 1530–81)

Biography in J. Pulver, *A Biographical Dictionary of Old English Music* (1927). Master of the Children of the Chapel Royal from 1576–81; see H. N. Hillebrand, *The Child Actors* (1926) and G. E. P. Arkwright, 'Elizabethan choirboy plays and their music', *Proceedings of the Musical Association*, 1913–14. Farrant is often supposed to be the author of the anonymous *The Wars of Cyrus* (*q.v.*).

FOXE, JOHN (1517–87)

Foxe's major works are discussed and listed in C. S. Lewis, OHEL. His one dramatic work, *Christus Triumphans: comoedia apocalyptica* was first published in Basel in 1556. It was acted in Cambridge 1562/63. The original manuscript is now in the British Museum (Lansdowne MS. 1073). It was translated into English by R. Day in 1579 (many subsequent editions) and into French by Jean Bienvenu (1562).

FRAUNCE, ABRAHAM (*fl.* 1587–1633)

See C. S. Lewis, OHEL, for Fraunce's non-dramatic writings. The one play certainly his, *Victoria*, a version of Luigi Pasqualigo's *Il Fedele* (1576), is preserved in manuscript. It is printed ed. G. C. Moore Smith in Materialien, 1906. See F. S. Boas, *University Drama*, and G. C. Moore Smith, 'Notes on some English University plays', *MLR*, 1908. For the anonymous *Hymenaeus*, sometimes attributed to Fraunce, see s.v.

FULWELL, ULPIAN (*fl.* 1568; d. 1586)

Irving Ribner (*NQ*, 1950) establishes what is known about the author. His *Ars Adulandi* is established (*NQ*, 1951) as a largely personal work aimed at the Chapter of Wells Cathedral. His one play, *Like will to Like*, S.R. 1568; pub. 1568; *c.* 1570; 1587, has been reprinted in H-D 3 and by J. S. Farmer, 1906, 1909*.

GAGER, WILLIAM (1555–1622)

The fullest account of Gager's life is in C. F. Tucker Brooke's 'The life and times of William Gager', *Proceedings of the American Philosophical Society*, 1951. F. B. Williams, Jr., writes in *TLS*, 18 April 1936, on 'Gager's will'. Gager's writings are discussed in Leicester Bradner, 'A bibliography of Neo-Latin drama', *Studies in the Renaissance*, 1954. See also F. S. Boas, *University Drama in the Tudor Age* (1914). There are synopses of the extant plays in *Sh. Jb.*, 1898.

*Meleager, tragoedia nova*, was acted in Christ Church, Oxford, in 1582, published 1592. It is described in Boas, op. cit., pp. 165–78. His *Rivales*, which was performed in 1583, has perished. What is known of it is summarized in Boas, pp. 181 et seq.

*Dido*, also performed in 1583, survives in two manuscripts. It was printed (in part) by A. Dyce in *The Works of Marlowe* (1850). It is described in Boas, pp. 183–9.

*Ulysses Redux* was performed in 1592 and published in the same year. It is described in Boas, pp. 201–19.

The *Panniculus Hippolyto Senecae Tragoediae assutus*—additions to a performance of Seneca's *Hippolytus* in 1592—was printed as an appendix to the *Meleager* (1592). It is described in Boas, pp. 198–201. Gager's *Oedipus*, of which a fragment survives in manuscript, is printed by R. H. Bowers in *SP*, 1949. For his Prologue and Epilogue to the anon. English version of *Bellum Grammaticale* see s.v.

Gager's Latin poems are described and some of them are printed by Tucker Brooke in *SP*, 1932. The same author in 'Latin Drama in Renaissance England' (*ELH*, 1946) draws heavily on Gager for his illustrations.

At the time of his death Tucker Brooke was working towards an edition of Gager's plays. His manuscript is deposited in the library of the American Philosophical Society.

Gager's controversy with Rainolds on the propriety of stage plays is reprinted in part by F. S. Boas (Chapter X). See also K. Young, 'William Gager's defence of the academic stage', *Transactions of the Wisconsin Academy*, 1916, W. Ringler, 'The first phase of the Elizabethan attack on the stage 1558–1578', *HLQ*, 1942, and E. N. S. Thompson, *The Controversy between the Puritans and the Stage* (1903).

GARTER, THOMAS (*fl.* 1569)

Useful biographical speculations appear in the preface to the MSR. *The Comedy Of the moste vertuous and Godlye Susanna*, written ?; S.R. 1563 ['a ballett of the godly and constante wyse Susanna'], 1569; published 1578.

MSR, 1937, edited by B. I. Evans and W. W. Greg.

See B. I. Evans in *TLS*, 1936 (p. 372) for an account of the discovery of the unique copy of this play.

Robert Pilger in 'Die dramatisierungen der Susanna im 16 Jahrhundert' (*Zeitschrift für deutsche philologie*, 1880) gives a detailed account of many versions of the Susanna story in this period; cf. M. T. Herrick, 'Susanna and the Elders in sixteenth century drama', *Studies in Honour of T. W. Baldwin* (1958).

GASCOIGNE, GEORGE (*c.* 1539–77)

C. T. Prouty's *George Gascoigne* (1942) is a complete account of what is known about the man. S. A. Tannenbaum has a *Concise Bibliography* of writings by and about Gascoigne (1942). See also C. S. Lewis in OHEL. Of his three plays, *Jocasta* and *Supposes* (1566) were printed in *A Hundreth sundrie floures* (1573)—ed. Prouty, 1942—in *The posies of George Gascoigne* (1575), in *The whole woorkes of George Gascoigne* (1587), in W. C. Hazlitt's *Complete poems of George Gascoigne* (1869–70), ed. J. W. Cunliffe (Belles Lettres Series; 1906), in J. W. Cunliffe's *Works of George Gascoigne* (1907–10). *Jocasta* has also been printed in F. J. Child, *Four Old Plays* (1848) and in J. W. Cunliffe's *Early English Classical Tragedies* (1912). *Supposes* appears in T. Hawkins's *The Origin of English Drama, III* (1773) and in R. W. Bond's *Early Plays from the Italian* (1911), in G. Bullough's *Narrative and Dramatic Sources of Shakespeare*, I (1957).

*The Glasse of Governement* (1575) is reprinted in *Complete Poems*, ed. W. C. Hazlitt (1869–70), and in *Works of George Gascoigne*, ed. Cunliffe (1907–10) and is ed. Farmer, 1914*.

*The Princely pleasures at Kenelworth Castle* (performed 1575) was first printed in 1576. The only copy surviving disappeared in 1879, but had been previously reprinted in anon., *Kenilworth Illustrated* (1821); it was reprinted in *The whole woorkes of George Gascoigne* (1587). The work is also printed in J. Nichols, *The Progresses of Queen Elizabeth* (1788–1821).

GOSSON, STEPHEN (1554–1624)

The definitive account of Gosson's life and works is W. Ringler, *Stephen Gosson* (1942). See also C. S. Lewis, OHEL. Three plays by Gosson are known only by name, being mentioned by Gosson himself and by Thomas Lodge, during their controversy. Two of them (*The Comedie of Captain Mario* and *Praise at Parting*) seem to be dated *c.* 1577; the third (*Catilin's Conspiracies*) seems to be 1578/79. Gosson's *School of Abuse* (1579; 1587), attacking the stage (*inter alia*) is reprinted in *The Somers Tracts* (edited Walter Scott), vol. iii (1810), edited J. P. Collier, Shakespeare Society (1841) and, together with his *Apologie of the School of Abuse* (1579, etc.), reprinted by E. Arber (1868). His *Plays confuted in five actions* (1582) is reprinted in W. C. Hazlitt, *The English Drama and Stage* (1869).

GRIMALD, NICHOLAS (1519?–1562?)

*The Life and Poems of Nicholas Grimald* by L. R. Merrill (1925) gives the most complete account of this author. It translates much of Grimald's Latin into English. On Grimald's non-dramatic poetry see C. S. Lewis (OHEL).

*Archipropheta* was written 1547?; published (Cologne) 1548. F. S. Boas, *University Drama*, describes it pp. 33–41.

*Christus Redivivus, comoedia tragica, sacra et nova* was written *c.* 1540; published (Cologne) 1543. The text is edited by J. M. Hart in *PMLA*, 1899. It is described in Boas, op. cit., pp. 26–32. Some attempts have been made to link this play with the medieval English drama; G. C. Taylor writes on 'The *Christus Redivivus* of Nicholas Grimald and the Hegge Resurrection Plays' (*PMLA*, 1926) and Patricia Abel on 'Grimald's *Christus Redivivus* and the Digby Resurrection Play', *MLN*, 1955.

HEYWOOD, JOHN (1497–1578)

Heywood's life, his association with More, Rastell, Medwall, etc., is discussed in A. W. Reed's *Early Tudor Drama* (1926), in R. W. Bolwell, *Life and Works of J. Heywood* (1921), in R. de la Bere, *John Heywood 'Entertainer'* (1937), and Pearl Hogrefe, *The Sir Thomas More Circle* (1959). There is a *Concise Bibliography* by S. A. and D. R. Tannenbaum (1946). For the canon of six plays usually assigned to him see Hillebrand in *MP*, 1915/16 and Reed, op. cit. (also *The Library*, 1918). Sources are discussed by Karl Young (*MP*, 1904/5)—corrected by both T. W. Craik (*MLR*, 1950) and by W. Elton (*TLS*, 1950, p. 128)—and by I. C. Maxwell, *French Farce and John Heywood* (1946). The chronology of the six plays is the subject of W. Phy in *E. Studien*, 1940. Farmer has edited the *Dramatic Works* (1905).

The least clearly authenticated play, *Johan Johan* (1533), is in Brandl, Gayley, Adams, Farmer (1909)*, etc., *The Play of Love* (1534) is in Brandl, Farmer (1909)*, and ed. K. W. Cameron (1944)—see also R. J. Schoeck in *NQ*, 1951, pp. 112 et seq. *The Pardoner and the Frere* (1533) is in H-D 1, Farmer (1909)*, and in Pollard's *English Miracle Plays* (1927). *The Four PP* (? 1544) is in H-D 1, Manly, Adams, Farmer (1908)*, and Boas, *Five Pre-Shakespearean Comedies* (1934). *The Play of the Wether* (1533) is in Brandl, Gayley, Adams, Farmer (1908)*, and edited K. W. Cameron (1941); the source is discussed in *MLN*, 1907. See also D. M. Bevington, 'Is John Heywood's *Play of the Weather* really about the weather?' *Renaissance Drama*, 1964. *Witty and Witless* (or *A Dialogue on Wit and Folly*) was first printed from the manuscript by F. W. Fairholt (Percy Society, 1846); it is printed in Bolwell's and de la Bere's books and edited by K. W. Cameron (1941).

INGELEND, THOMAS (*fl.* 1560)

Only known for certain as the author of *The Disobedient Child*, possibly written about 1560 and printed *c.* 1570; ed. J. O. Halliwell, (Percy Soc., 1848); H-D 2; Farmer, 1905, 1908*. F. Holthausen printed in *E. Studien*, 1902, the dialogue of Ravisius Textor ('Iuvenis, Pater, Uxor') on which part of the play is based—a source first pointed out by Creizenach.

JEFFERE, JOHN

Known only as the author of *The Bugbears*, a translation (? before 1566) of A. Grazzini, *La spiritata* (1561), preserved in manuscript. It was printed (with Quellenuntersuchung) by C. Grabau, *Archiv*, 1897, and by R. W. Bond, *Early Plays from the Italian* (1911).

KINWELWERSHE, FRANCIS (*fl.* 1557–80?)

For *Jocasta*, see s.v. Gascoigne.

LEGGE, DR. THOMAS (1535–1607)

Master of Caius College, Cambridge. His Latin play *Richardus Tertius* was acted 1579 at St. John's College, Cambridge. It is extant in eleven manuscripts. A summary appears in G. B. Churchill and W. Keller, *Sh. Jb.*, 1898. It has been reprinted for the Shakespeare Society, 1844, and in the Hazlitt–Collier, *Shakespeare's Library*, vol. v (1875). See G. B. Churchill, *Richard the third up to Shakespeare* (Berlin, 1900).

LINDSAY (LYNDSAY), SIR DAVID (1490?–1555)

For works other than *Ane Satyre of the Thrie Estaits* (acted in various versions 1539–1554?) see C. S. Lewis (OHEL). There is a bibliography in *The Library*, 1929. The Bannatyne MS. (which contains several sections of the play) is in STS (ed. Ritchie). The full version (Edinburgh, 1602) appears in the modern editions of the *Works* of Lindsay—together with the Bannatyne sections in the STS edition (ed. Hamer). There is a version abridged by J. Kinsley (1954); and modernized abridged versions by R. Kemp—as presented at the Edinburgh Festival—(1951), and by M. P. McDiarmid (1967). The text is further discussed in *PMLA*, 1940. Anna J. Mill gives an account of the original performances in *PMLA*, 1932 (with corrections in *PMLA*, 1933). She discusses foreign influence in *MLR*, 1930. See also W. Murison, *Sir David Lyndsay, Poet and Satirist* (1938), and Craik, pp. 93 ff., etc.

LUPTON, THOMAS (*fl.* 1583)

A miscellaneous author, who appears in this list because of his *moral and pitieful Comedie, Intituled All for Money* (1578), (S.R. 1577), ed. J. O. Halliwell in *Literature of the Sixteenth and Seventeenth Centuries Illustrated* (1851), in *Sh. Jb.*, 1904 and ed. J. S. Farmer, 1910*. T. W. Craik (*NQ*, June 1954) has noted a borrowing from one of Latimer's sermons. See Bevington, pp. 165 ff., etc.

Lupton's other works are *A thousand notable things of Sundrie sorts* (1579, etc.); *SIVQILA. Too good to be true . . . the wonderful maners of the people of Mauqsun* (1580, etc.); *The second part and knitting up of the Boke entituled, Too good to be true* (1581); *A persuasion from Papistrie* (1581); *The Christian against the Jesuite* (1582); *A dreame of the devill and Dives* (1584).

MEDWALL, HENRY (born *c.* 1462)

The best account of Medwall's work appears in A. W. Reed's *Early Tudor Drama* (1926); see also Pearl Hogrefe, *The Sir Thomas More Circle* (1959); some basic dates are provided by Sir Wasey Sterry in *The Times* 15 April 1936. His morality-play *Nature* (? 1516–20) is printed in Brandl, in Farmer, 1907 and 1908. A fragment is in Materialien edn. of *Youth* (1905). W. R. Mackenzie (*PMLA*), 1914, suggests that it is derived from Lydgate's *Reson and Sensuallyte*. His *Fulgens & Lucres* (written before 1500; published *c.* 1515) was known to Halliwell-Phillips in 1885 (*Outlines*, ii. 340) but seemed to have survived only in a fragment, which was printed as *Lucrece* ed. Bang and McKerrow, Materialien, 1905, and ed. Greg, MSC, 1 (ii), 1909. Creizenach noted in *Sh. Jb.*, 1911, that the source of the fragment was to be found in Bonaccorso's *De Vera Nobilitate*, and Reed (op. cit.) later pointed to Tiptoft's translation of this as the immediate source. The discovery of the full play in the Mostyn sale of 1919 was announced by Boas (*TLS*, 20 February 1919) and its purchase by Henry E. Huntington was followed by a facsimile reprint by de Ricci (1920), and editions by Boas and Reed (1926), and Boas (*Five Pre-Shakespearean Comedies*, 1934). Medwall's debate technique has been discussed by E. M. Waith (*PMLA*, 1953) and his conventions by C. R. Baskervill (*MP*, 1927). See also J. K. Lowers, 'High comedy elements in Medwall's *Fulgens & Lucres*' (*ELH*, 1941); C. E. Jones, 'Notes on *Fulgens and Lucres*', *MLN*, 1935. Medwall's *The Fyndyng of Troth* which was described by Collier in his *History of Dramatic Poetry* (p. 69) may be Collier's invention.

NORTON, THOMAS (1532–84)

There is an extensive account of Norton's life (by Sidney Lee) in *DNB*.

The *Tragedie of Gorboduc* ('whereof three Actes were wrytten by Thomas Nortone, and the two laste by Thomas Sackvyle') was acted in the Inner Temple, Christmas 1561/2; entered in the Stationers' Register *c.* September 1565; published 1565 and n.d. [*c.* 1570] (*The Tragidie of Ferrex and Porrex*). Reprinted Dodsley, edited L. T. Smith, 1883, Manly II, edited Farmer 1906, 1908*; edited H. A. Watt, 1910, edited J. W. Cunliffe (*Early English Classical Tragedies*, 1912). The two quartos are compared by I. B. Cauthen, Jnr., in *Studies in Bibliography*, 1962. R. Y. Turner writes on 'Pathos and the *Gorboduc* tradition', *HLQ*, 1962. Senecan influence on *Gorboduc* is discussed by H. Schmidt (*MLN*, 1887); by Cunliffe, Baker, etc., cited under 'Seneca' in Section III; by M. T. Herrick in *Studies . . . in Honor of Alexander M. Drummond* (1944), and by P. Bacquet, *Études anglaises*, 1961. The political meaning of *Gorboduc* is discussed by S. A. Small (*PMLA*, 1931), by S. R. Watson, '*Gorboduc* and the theory of Tyrannicide', *MLR*, 1939, and by Gertrude Reese, ('The question of the Succession in Elizabethan Drama', *University of Texas Studies in English*, 1942).

Norton translated Calvin's *Institutes* (1561) and twenty-eight psalms in the Sternhold and Hopkins Psalter (1561).

PHILLIP, JOHN (*fl.* 1560–90)

*DNB* and *STC* seek to distinguish John Philip (*fl.* 1566) from John Phillips (*fl.* 1570–91); but W. W. Greg, *The Library*, 3rd series, 1910, asserts that the two entries refer to one man. The one play involved is the *Commodye of pacient and meeke Grissill*, written *c.* 1566; S.R. 1566 and 1569; published n.d. [? 1569]; MSR 1909.

Louis B. Wright (*RES*, 1928) suggests a political reference to Elizabeth's potential marriage. See Spivack, pp. 272 ff., etc.

Under the name John Philip appears an account of witch-trials in 1566. Under that of John Phillips appear twelve other patriotic and moralizing pamphlets—see *STC*.

PICKERING (PYKERYNG), JOHN (*fl.* 1567)

Only known from the one production that bears his name—*A Newe Enterlude of Vice Conteyninge the Historye of Horestes . . .* (1567). F. Brie (*E. Studien*, 1912) has listed several John Pickerings, but quite indecisively. J. E. Phillips in *HLQ*, 1954/55, picks up the suggestion in Carl Kipka's *Maria Stuart im Drama der Weltliteratur* (Leipzig, 1907) that the play is designed to reflect on Mary's reign, and that the author was (Sir) John Puckering (1544–96), ardent anti-Marian and subsequent Lord Keeper. The play is probably the same as the *Orestes* acted at Court in 1567–8. It is printed in J. P. Collier, *Illustrations of Old English Literature* ii (1866), in Brandl, Farmer (1910)*, MSR, 1962. Discussion appears in Willard Farnham, *The Medieval Heritage of Elizabethan Tragedy* (1936), and E. B. de Chickera, 'Horestes'

Revenge—another interpretation', *NQ*, May 1959. See also Spivack, pp. 279 ff., etc., Bevington, pp. 85 ff., etc.

PRESTON, THOMAS (?1537–98)

It has been doubted if the author of *Cambises* is the same as the Master of Trinity Hall, who was praised for his acting by Elizabeth when she visited Cambridge in 1564 (see Nichols, *Progresses of Queen Elizabeth*, I, 181). *Cambises* was ent. S.R. 1569; pub. (1) n.d. [1569?]; (2) n.d. [1585?]; (3) n.d. [1588?]. It is reprinted by T. Hawkins (1773), H-D 4, Farmer (1910)*, Adams, Manly. W. A. Armstrong (*E. Studien*, 1950) finds the source in Taverner's *Garden of Wisdom*; elsewhere (*E. Studies*, 1955) he finds covert political references. See A. Lincke, 'Kambyses in der Sage, Literatur und Kunst, des Mittelalters', *Aegyptiaca. Festschrift für Georg Ebers* (Leipzig, 1897); M. C. Linthicum, 'The date of *Cambyses*', *PMLA*, 1934; M. P. Tilley, 'Shakespeare and his ridicule of *Cambyses*', *MLN*, 1909; Spivack, pp. 284 ff., etc.; Bevington, pp. 183 ff., etc.

RASTELL, JOHN (1475?–1536)

Rastell's life as printer, impresario, merchant-venturer, and author is discussed at length in A. W. Reed's *Early Tudor Drama*.

His authorship of *The Nature of the Four Elements* (1517–27), which in the sole copy surviving lacks colophon and the last part of the text and some of the middle, is attested by Bale. It is printed by J. O. Halliwell (Percy Society, 1848), in H-D 1, ed. J. Fischer in *Marburger Studien*, 1903, by Farmer in 1905 and 1908*, and in A. W. Pollard's *English Miracle Plays* (1927). The nature and the sources of Rastell's geographical knowledge in this play have been canvassed in *SP*, 1938, *PQ*, 1938, *PMLA*, 1942, 1943, 1945. See also P. Hogrefe, *The Sir Thomas More Circle* (1959), and Spivack, pp. 86 ff., etc.

For *Calisto and Melebea* and *Gentylnes and Nobylyte*, often attributed to Rastell, see under these titles in 'Anonymous Plays' section.

For his chronicle history, *Pastyme of People*, see A. W. Reed, *Early Tudor Drama*, Appendix IV.

REDFORD, JOHN (*c.* 1486–1547)

Master of the Children of St. Paul's 1536–47. Biography in W. H. Grattan Flood, *Early Tudor Composers* (1925) and

J. Pulver, *A Biographical Dictionary of Old English Music* (1927). See Grove (1954), H. N. Hillebrand, *The Child Actors* (1926), C. Pfatteicher, *John Redford* (Kassel, 1934), Pearl Hogrefe, *The Sir Thomas More Circle* (1959), and A. Brown, 'Two Notes on John Redford', *MLR* 1948.

His *Wit and Science* (MS.—written *c.* 1530?) was edited by Halliwell (Shakespeare Society, 1848); Manly I; Farmer, 1907; 1908*; Adams, MSR 1951.

See J. Siefert, *Wit-und-Science Moralitäten* (Prague, 1892); H. Hauke, *John Redford's Moral Play 'The Play of Wit and Science' und seine spätere Bearbeitung* (1904); S. A. Tannenbaum, 'Editorial notes on *Wyt and Science*', *PQ*, 1935; A. Brown, 'The Play of *Wit and Science*', *PQ*, 1949; W. Habicht, 'The Wit-Interludes and the form of pre-Shakespearean "Romantic Comedy" ', *Renaissance Drama*, 1965.

MR. S. MR OF ART (*fl. c.* 1553)

The author of *Gammer Gurtons Nedle* was identified as William Stevenson (Fellow of Christ's College, Cambridge) by Henry Bradley (*Athenaeum*, 6 August 1898, and Gayley I). An earlier identification with John Still (later Bishop of Bath and Wells) was conjectured by Isaac Reed in the *Biographia Dramatica* II (1782).

The author of two Marprelate tracts had assumed that John Bridges (later Bishop of Oxford) was the author, and this has been supported by C. H. Ross (*MLN*, 1892, and *Anglia*, 1897). C. W. Roberts (*PQ*, 1939) has sought to give the comedy to Sebastian Westcott (q.v.).

The fullest account of the authorship controversy is in F. S. Boas, *University Drama* (1914). The play was written *c.* 1553?; entered S.R. (as 'Dycon of Bedlam') in *c.* January 1563; the first extant edition is 1575; but there may have been an earlier one. Reprinted (e.g.) 1661, Dodsley, Manly II, Gayley I, Farmer, 1906, 1910*, ed. H. Brett-Smith, 1920, ed. F. S. Boas (*Five Pre-Shakespearean Comedies*, 1934). See also B. J. Whiting, 'Diccon's French Cousin', *SP*, 1945; H. A. Watt, 'The Staging of *Gammer Gurtons Needle*', *Elizabethan Studies . . . in Honor of G. F. Reynolds* (1945); Spivack, pp. 322 ff., etc.

SACKVILLE, THOMAS (1536–1608)

For non-dramatic works see C. S. Lewis, OHEL. For *Gorboduc* see s.v. Norton. The latest monograph is P. Bacquet,

*Un Contemporain d'Elisabeth I: Thomas Sackville, l'homme et l'œuvre* (Geneva, 1966).

SKELTON, JOHN (1460?–1529)

For non-dramatic works see Lewis, OHEL. *Magnyfycence* (*c.* 1515) has been edited separately by R. L. Ramsay for EETS (ES 98, 1908) and by Farmer, 1910*, and is printed in Skelton's *Works* ed. Dyce (2 vols., 1843) and P. Henderson (1931). See also A. R. Heiserman, *Skelton and Satire* (1961), Bevington (pp. 132 ff., etc.) and W. O. Harris, *Skelton's Magnyfycence and the Cardinal Virtue Tradition* (1965).

TARLTON, RICHARD (*d.* 1588)
clown, actor, playwright

His *Seven Deadly Sins* (? 1585)—also known, perhaps, as *Five* [*Four, Three*] *Plays in One*—has not survived; but the *Secounde Parte of the Seven Deadlie Sinns* has its 'platt' or scenario preserved in manuscript at Dulwich College. It is reprinted in the Variorum Shakespeares of 1803 and 1821 and in W. W. Greg, *Elizabethan Dramatic Documents* (1931). Other works involving Tarlton's name (such as *Tarltons Newes out of Purgatorie*, 1590; *Tarltons jests*, 1638) are not necessarily written by Tarlton.

UDALL, NICHOLAS (*c.* 1505–56)

For Udall's life see *DNB*, supplemented by (e.g.) E. Flügel's essay in Gayley I, A. W. Reed, *RES*, 1925; Materialien, 1939.

*Ralph Roister Doister* is known to be Udall's because of a reference in Thomas Wilson's *Rule of Reason* (1553 edition): written 1545–53, it was entered in S.R., *c.* October 1566, and published n.d. [*c.* 1566?]. It is reprinted (e.g.) ed. E. Arber 1869, H-D 3, Manly II, Gayley I, Adams, MSR, 1935, ed. Boas (*Five Pre-Shakespearean Comedies*, 1934), Materialien, 1939, J. S. Farmer has printed *The Dramatic Writings of Udall* (1906).

On the date of *Ralph Roister Doister* see J. W. Hales, *E. Studien* 1893, S. Gaselee, *Etoniana*, 28 December 1943; T. W. Baldwin and M. C. Linthicum, *PQ*, 1927. See also T. W. Baldwin, *William Shakspere's Five-Act Structure* (1947), W. L. Edgerton, *PQ*, 1965, Spivack, pp. 318 ff., etc.

On the sources see D. L. Maulsby in *E. Studien*, 1907 and J. Hinton *MP*, 1913. See also E. S. Miller, 'Roister Doister's "Funeralls" ', *SP*, 1946.

Udall's lost *Ezechias* is discussed by A. R. Moon, *TLS*, 19 April 1928. For arguments that Udall wrote other plays see s.v. *Respublica, Jacke Jugeler, Jacob and Esau, Thersites.*

Udall's other established works are mostly translations from Latin for religious or pedagogic uses: his poems are printed in J. Nichols, *Progresses of Queen Elizabeth* I (1788); ed. Furnivall, *Ballads from Manuscripts* (1870); ed. Arber, *English Garner* II (1879).

WAGER, (?) (*fl.* 1566)

*The cruel debtor*, which may be by either Wager (the Stationers' Register enters it to 'Wager' in 1566) has only four leaves surviving. They were printed by F. J. Furnivall, *New Shakspere Society Transactions*, 1877–9 (three leaves only) and ed. W. W. Greg, MSC, I (iv and v), 1911 and MSC, II (ii), 1923.

WAGER, LEWIS (*fl.* 1566)

Wager's *New enterlude ... entreating of the life and repentaunce of Marie Magdalene*, published 1566; 1567, was entered in the Stationers' Register December 1566/January 1567. It was edited by F. I. Carpenter, 1902; revised edition 1904. Farmer, 1908*. There is an important review of Carpenter's edition by R. Imelman, *Archiv*, 1903. See Spivack, pp. 262 ff., etc., Bevington, 94 ff., etc.

WAGER, WILLIAM (*fl.* 1565–9)

On the confusion of identity between the 'Wagers' see Greg in MSC, I (iv and v), 1911, pp. 324–7.

*Inough is as good as a feast* (written 1564?; published n.d. [1565?]) has been reprinted by S. de Ricci, 1920*. See Spivack, pp. 171 ff., etc., Craik, pp. 99 ff., etc., Bevington, pp. 158 ff., etc.

*The longer thou livest, the more foole thou art* (written 1564?; S.R. 1569; pub. n.d. [1569?]) was edited A. Brandl, *Sh. Jb.*, 1900, Farmer, 1910*. For *The Trial of Treasure* see under 'Anonymous Plays'. See Bevington, pp. 163 ff., etc., 153 ff., etc., Craik, pp. 57 ff., etc.

WAPULL, GEORGE (*fl.* 1576)

*The Tyde taryeth no Man* (written 1576?; Stationers' Register, October 1576; published 1576) has been edited by J. P. Collier

(*Illustrations of . . . Popular Literature*, 1863) and by E. Rühl, *Sh. Jb.*, 1907, Farmer, 1910*. See also Bevington, pp. 149 ff., etc.

WATSON, THOMAS (*c.* 1515–84)

For his writings see *STC*. His Latin play *Absalom* is discussed in F. S. Boas, *University Drama*, and summarized by Churchill and Keller in *Sh. Jb.*, 1898; it has been edited and translated by J. H. Smith (1964).

WESTCOTT, SEBASTIAN (*d.* 1582)

Westcott's work as Master of the St. Paul's boys is described in H. N. Hillebrand's *The Child Actors* (1926; 1946). A. Brown has supplemented what we know in 'Sebastian Westcott at York', *MLR*, 1952 and 'Three notes on Sebastian Westcott', *MLR*, 1949. It is sometimes supposed that some of the many plays Westcott presented at court were of his authorship. J. P. Brawner, 'Early Classical Narrative plays by Sebastian Westcott and Richard Mulcaster', *MLQ*, 1943, is corrected by A. Brown 'A note on Sebastian Westcott and the plays presented by the Children of Paul's', *MLQ*, 1951, which also picks up the earlier assumption of H. N. Hillebrand that Westcott was the author of *Liberality and Prodigality* (q.v.).

WEVER, RICHARD (*fl. c.* 1549–53)

Known only as the author of *Lusty Juventus*, written *c.* 1550?; ent. S.R., 1560; pub. (1) n.d. [? 1565]; (2) n.d. [? 1565]; (3) n.d. [? 1565]. Ed. T. Hawkins, *Origins of the English Drama*, vol. i (1773); H-D 2, Farmer, 1905, 1907*. See Bevington, pp. 143 ff., etc.

WHETSTONE, GEORGE (1544/51–1587)

The standard work is T. Izard, *George Whetstone: Mid-Elizabethan Gentleman of Letters* (1942). His one play *Promos and Cassandra*, printed in 1578, is reprinted in [J. Nichols] *Six Old Plays* (1779), Collier-Hazlitt, *Shakespeare's Library*, VI (1875), Farmer, 1910*, G. Bullough, *Narrative and Dramatic Sources of Shakespeare*, II (1958). The prose version of the same story, which Whetstone published in *Heptameron of Civill Discourses* (1582), is reprinted in Collier-Hazlitt's *Shakespeare's Library*, III (1875). See F. E. Budd, 'Rouillet's *Philanira* and Whetstone's *Promos* *and Cassandra*', *RES*, 1930, and the same author, 'Material for a study of the sources of Shakespeare's *Measure for Measure*', *Revue de Littérature Comparée*, 1931. See also C. S. Lewis, OHEL.

WILMOT, ROBERT, (*fl.* 1568–1608)

*DNB* has a life of Wilmot, not notably out-of-date. *Tancred and Gismund* was published by Wilmot in 1591/1592, as the work of Robert Stafford, Henry Noel, G. Al., Christopher Hatton, and himself ('R.W.'). An earlier version, *Gismond of Salerne*, is extant in two manuscripts, printed (and collated) in Brandl and J. W. Cunliffe, *Early English Classical Tragedies* (1912), also presented by Farmer, 1912. Wilmot's revision is reprinted in H-D 7, Farmer 1912*, MSR, 1914. A. Klein, 'The decorum of these days', *PMLA*, 1918, compares the versions. J. W. Cunliffe, *PMLA*, 1906, describes the borrowings; J. Murray, *RES*, 1938, discusses the general problems of the plays. See also W. Habicht 'Die Nutrix-Szenen in *Gismond of Salern* und *Tancred and Gismund*', *Anglia*, 1963.

WILSON, ROBERT (R.W.) (*fl.* 1572–1600)

It is sometimes supposed (e.g. by Sidney Lee in *DNB*) that there were two Robert Wilsons, one an actor mentioned from 1572 as belonging to Leicester's players, and the author of plays mentioned below, the other his son, born in 1579, mentioned by Meres in 1598 and by Henslowe's diary as collaborating in various plays, of which only one, *Sir John Oldcastle*, survives. I. Gourvitch (*NQ*, 1926, pp. 4–6) argues for one man, and Chambers supports this (*ES*, ii. 349 f.). The two morality plays, *The three ladies of London* (1584; 1592) and *The three lordes and three ladies of London* (1590) are attributed only to 'R.W.' but this is regularly supposed to be Wilson. *Three Ladies* is edited by J. P. Collier (*Five Old Plays*, 1851), in H-D 6, Farmer 1911*. *The three lordes and three ladies* appears in the same reprints (Farmer, 1912*). Miss I. Mann (*PMLA*, 1944) thinks the latter play to be abridged. *The Coblers Prophesie* (1594) is attributed to Robert Wilson, Gent. It is edited in *Sh. Jb.*, 1897, Farmer 1911*, MSR, 1914. On the bibliography of the play see Miss I. Mann, *Library* (fourth series) 1943, and 1946, *NQ*, 1945 (pp. 48–50). One further play, *The Pedlers Prophecie* (1595) has been attributed to Wilson on the basis of the analogy with *The Coblers Prophesie*. See s.v. under 'Anonymous Plays'.

WOODES, NATHANIEL (*c.* 1550– after 1594)

The little that is known of Woodes's life appears in Celesta Wine's article (*RES*, 1939). The *Conflict of Conscience* was published in 1581 in two states; one has Francis Spera's name on the title-page and a tragical conclusion; the other calls him *Philologus* and has a happy ending—see *TLS*, 1933, p. 592 and p. 732. It has been reprinted ed. Collier (Roxburghe Club, 1851), H-D 6, Farmer (1911*) and MSR, 1952. L. M. Oliver, *RES*, 1949, suggests Foxe as a source; *PMLA*, 1935, discusses the sources very fully. L. B. Campbell in *PMLA*, 1952, explores the relationship of the play to Marlowe's *Dr. Faustus*.

## V. ANONYMOUS PLAYS

*Albion Knight.* Written 1537?–65; S.R. 1565; published 1566? A fragment survives, edited by Collier (1844), Farmer (1906), and by Greg (MSC, I (iii), 1909). It has been thought by A. G. Jones to refer to Parliamentary difficulties in the middle of Elizabeth's reign (*JEGP*, 1919) and referred (by M. H. Dodds) to the Pilgrimage of Grace (*The Library*, 1913).

*Bellum Grammaticale.* Dramatised version probably by Leonard Hutten (? 1557–1632), and probably written *c.* 1583; was acted at Christ Church in 1592 (with prologue and epilogue by William Gager), and printed in 1635. An English adaptation [? by S. Hoadley], acted 1666, is in BM Add. MS. 22725. See J. Bolte, *Andrea Guarnas, Bellum Grammaticale und seine Nachahmungen* (Berlin, 1908).

*Calisto and Melebea.* ('A new comodye in englysh in maner of an enterlude ryght elygant and full of craft of rethoryk wherein is shewd and dyscrybyd as well the bewte and good propertes of women as theyr vycys and evyll condicions') was probably written and printed *c.* 1527. It is often attributed to John Rastell. It is printed in H-D 1, Farmer (1905, 1909*), and MSR, 1908. See A. W. Reed, *Early Tudor Drama*, and A. S. V. Rosenbach (*Sh. Jb.*, 1903).

*Clyomon and Clamydes.* Written *c.* 1570–83, published 1599; attributed to Peele by Dyce, and printed in the editions of Peele by Dyce (1839, 1861), and by Bullen (1888). Printed also by Farmer (1913*) and MSR, 1913. Kellner, *E. Studien*, 1889, has a full discussion of the play, and is inclined to allow Peele's authorship. Kittredge, *JEGP*, 1898, suggests Preston as the author. In this century it seems to have regained its status as a play of unknown authorship. See Ellison, *Romantic Drama*, for discussion of the source. See also Spivack, pp. 299 ff., etc., and Bevington, pp. 194 ff., etc.

*Common Conditions*. Written and published *c.* 1576 (two edns.); ed. Tucker Brooke (1915)—ed. (1); Brandl, Farmer (1908)—ed. (2). Tucker Brooke (*MLN*, 1916) picks up the suggestion of Marie Gothein (*Sh. Jb.*, 1904), and shows that the *Amor Costante* of Piccolomini is an interesting analogue. See Spivack, pp. 291 ff., etc., and Bevington, pp. 191 ff., etc.

*The Contention between Liberality and Prodigality*. See under *Liberality*.

*The Contract of a Marriage between Wit and Wisdom*. See under *Marriage*.

*Fedele and Fortunio* (translation of *Il Fedele* by Luigi Pasqualigo)—sometimes assigned to Anthony Munday. S.R. November 1584; published 1585; ed. Flügge (*Archiv*, 1909), MSR, 1909.

*Four Cardinal Virtues*. A fragment of an interlude written 1537–47 and printed by W. Middleton, 1541–7. It has been edited by W. W. Greg in MSC, IV (1956), who suggests a connexion between this fragment and *Temperance and Humility* (q.v.). See also W. O. Harris, *Skelton's Magnyfycence and the Cardinal Virtue Tradition* (1965).

*Gentylnes and Nobylyte*. Written *c.* 1527; printed (1) n.d. [? 1527–9] and (2) n.d. [? 1535]. It has been edited by E. Brydges and J. Haslewood, *The British Bibliographer*, vol. iv (1814), by J. H. Burn, 1829, J. S. Farmer, 1908 and 1908*, K. W. Cameron, 1941, MSR, 1950. The authorship is usually attributed to Rastell (A. W. Reed, *Early Tudor Drama*, E. C. Dunn, *MLR*, 1917) or Heywood (Tucker Brooke, *MLR*, 1911, and K. W Cameron, *Authorship and Sources of 'Gentleness and Nobility'* 1941).

*Godly Queene Hester*. Written *c.* 1527; S.R. 1561; pub. 1561. Ed. by J. P. Collier, *Early English Popular Literature*, vol. i (1863); A. B. Grosart, *The Fuller Worthies' Library Miscellanies*, iv (1873); W. W. Greg, Materialien, 1904; J. S. Farmer, 1906. See Spivack, pp. 256 ff., etc.

*Good Order*. See s.v. *Old Christmas or Good Order*.

*Hycke Scorner*. Written *c*. 1513; published (1) n.d. [? *c*. 1516] (2) fragment n.d. [1526–7]; (3) n.d. [? 1550]. Edited by T. Hawkins, *Origin of English Drama* I, 1773, H-D 1, Manly I, Farmer 1905, 1908*. See also under *Youth*.

*Hymenaeus*. Acted at St. John's College, Cambridge, 1579, and preserved in two Cambridge manuscripts; first printed, ed. G. C. Moore Smith, 1908; synopsis in *Sh. Jb.*, 1898, together with discussion of source (*Decameron* iv. 10), etc. The authorship has been attributed to Henry Hickman, and to Abraham Fraunce, whose candidature has been supported by F. S. Boas (*University Drama*).

*Impacyente Poverte*. Written 1547–58; entered S.R. 10 June 1560; published 1560. Ed. Farmer, 1907*, and Materialien, 1911. See Bevington, pp. 141 ff., etc.

*Jacke Jugeler*. Written ? 1553; S.R. 1562; published (1) n.d. [? 1563]; (2) n.d. [? 1565]; (3) n.d. [? 1565–70]. Ed. J. Warwick, 1820; F. J. Child (*Four Old Plays*), 1848; A. B. Grosart, *Fuller Worthies' Misc.* iv (1873), H-D 2; Ashbee, 1876*, Farmer, 1906, 1912*; ed. W. H. Williams, 1914, MSR, 1933 (first edition), MSR, 1937 (third edition). The authorship of this play has been attributed to Udall. W. H. Williams (*MLR*, 1912) argues for a date *c*. 1542 and posits parallels with Udall's other works. G. Dudok (*Neophilologus*, 1916) carries on the same argument, by citation of further parallels. C. R. Baskervill, reviewing Williams' edition, *MP*, 1916/17, finds little strength in his arguments. Gayley, *A Historical View of . . . English Comedy* (Gayley I) finds a concealed attack on transubstantiation. Variations from source are discussed by R. Marienstras in *Études anglaises*, 1963. See also Spivack, pp. 315 ff., etc.

*Jacob and Esau*. Written ? 1554; S.R. 1557; published 1568; in H-D 2; Farmer, 1906, 1908*; MSR, 1956. Authorship has been attributed to Udall (C. W. Wallace, *Evolution of English Drama*, and Leicester Bradner, *MLN*, 1927) and Hunnis (C. C. Stopes, *Athenaeum*, 28 April 1900, and *William Hunnis*). G. Scheurweghs, *E. Studies*, 1933, finds parallels with 1539 edition of Calvin's *Institutes*.

*Johan the Evangelyst*. Written 1520?; published n.d. [1550?]. Ed. MSR, 1907, Farmer, 1907*. Henry Bradley (*MLR* 1906/7) takes issue with Greg's MSR assumptions. W. H. Williams (*MLR* 1907/8) takes up some of Bradley's assumptions.

*Juli and Julian*. Written *c*. 1550–71; MS. in Folger Library. Ed. MSR 1955.

*Kyng Daryus*. Written 1565?; S.R. 1565; published (1) 1565; (2) 1577. Ed. J. O. Halliwell, 1860; Brandl; Farmer 1906, 1907* (edn. of 1577), 1909* (edn. of 1565). See Bevington, pp. 175 ff., etc.

*Liberality and Prodigality*. Written *c*. 1567–8; ? acted 1600; published 1602. Edited H-D 8, Farmer, 1912*, MSR, 1913. H. N. Hillebrand (*JEGP*, 1915) argues for the authorship of Sebastian Westcott; but A. Brown (*MLQ*, 1951), points to the danger of assuming that the master of a boys' company necessarily wrote their plays. See Craik, pp. 110 ff., etc.

*Love feigned and unfeigned*.—MS. fragment *c*. 1540–60; ed. A. Esdaile, MSC I (i), 1908. L. L. Scragg, *ELN*, 1966, discusses its technique of presentation.

*Mariage of Witte and Science*. Written 1567; S.R. *c*. August 1569; published n.d. [1569/70]. Fleay (*Chronicle* II, 288) and Harbage (*Annals*) identify with *Wit and Will*, which was played at Court in 1567/68 [Feuillerat, *Elizabeth*, p. 119]. See J. Siefert *Wit-und-Science Moralitäten des 16 Jahrhunderts* (Prague, 1892). Editions are H-D 2, Farmer, 1908, 1909*, MSR, 1961. See Collier's *History* (1879), II, 257–62. R. Withington (*PMLA*, 1942) discusses the meaning of *Experience* in the play. Sometimes thought to be an adaptation of Redford's *Wit and Science* (q.v.).

*The Contract of a Marriage between Wit and Wisdom*. Manuscript dated 1579 (or 1570) with the name of Francis Merbury appended to the Epilogue. Merbury may be the author. Greg (*Bibliography*, ii. 963–4) suggests that the manuscript may be a transcript of a lost printed book. Ed. J. O. Halliwell (Shakespeare Society, 1846), Farmer, 1908, 1909*. Greg and S. A. Tannenbaum have disputed the reading of the final figure in the manuscript date—Greg (*PQ*, 1932) defending '1579' and

Tannenbaum (*PQ*, 1930, *PQ*, 1933) defending '1570'. M. P. Tilley writes some 'Notes on the *Marriage of Wit and Wisdom*', *Shakespeare Assoc. Bulletin*, 1935.

*Minds* (*'a worke in ryme, contayning an Enterlude of Myndes'*), n.d. [*c.* 1574?]. From the low-German of Henrick Niklaes, founder of the Family of Love. Never reprinted [probably printed abroad].

*Misogonus*. Manuscript formerly at Chatsworth, now in Huntington Library, dated 1577. The name of Thomas Rychardes (written after the Prologue) was taken to be that of the author by Collier (*History*), Brandl, and Farmer. The title-page has the name Laurentius Bariωna, which R. W. Bond (*Early Plays from the Italian*, 1911) identifies as a real person, supposed to be the corrector rather than the author. G. L. Kittredge in *JEGP* 1901 had taken the name to be a quasi-Hebrew form for Laurence Johnson, the Catholic martyr, hanged 1582. S. A. Tannenbaum (*MLN*, 1930) identified another name on the title-page as not the unknown 'Anthony Rice', but 'Anthony Rudd' who was in Cambridge *c.* 1570; and G. C. Moore Smith (*TLS*, 10 July 1930) followed this lead; but Tannenbaum (*Shaksperian Scraps*, 1933) and D. M. Bevington (*ELN*, 1964) return to Johnson as the best candidate. Probably acted at Trinity College, Cambridge, *c.* 1568–74.

The play is edited by Brandl, Farmer (1906), R. W. Bond (*Early Plays from the Italian*, 1911). See also Spivack, pp. 327 ff., etc.

*Mundus et Infans*. See *The World and the Child*.

*New Custom*. Written 1570?; published 1573. Edited H-D 3, Farmer, 1906, 1908*. L. M. Oliver in *HLQ* 1947 notes that passages come from the 1570 edition of Foxe's *Acts and Monuments*. See Bevington, pp. 146 ff., etc.

*Nice Wanton*. Written 1547–53; S.R. 10 June 1560; published (1) 1560, (2) n.d. [1565?]. Brandl, p. lxxii suggests that the play adapts Macropedius's *Rebelles* (1535); edited H-D 2, Manly I, Farmer, 1905, 1909* (1560 edn.), 1908* (undated edn.).

*Old Christmas or Good Order*. Fragment of 131 lines and colophon, printed 1533. See G. L. Frost and R. Nash, *SP*, 1944.

*Pater, Filius, Uxor* [or 'The Prodigal Son']. A single leaf (? 1530–4) surviving from an interlude probably based on Textor's dialogue, *Iuvenis, Pater et Uxor* (also the source of Ingelend's *Disobedient Child*). Printed (as 'The Prodigal Son') in MSC I (i), 1908.

*Pedantius.* Latin comedy printed in 1631, and also extant in two Cambridge MSS., probably acted in Trinity College, Cambridge in 1581. Printed in Materialien, ed. G. C. Moore Smith, 1905. See H. S. Wilson on *Pedantius* and Gabriel Harvey, *SP*, 1948.

*The Pedlers Prophecie.* Written 1561?; S.R. May 1594; published 1595. Ed. Farmer, 1911* and MSR, 1914. G. L. Kittredge (*Harvard Studies and Notes in Philology*, 1934) points to 1561 as a probable date. The play is often attributed to Robert Wilson (q.v.).

*Processus Satanae* (the part of God in a play *c.* 1570). Printed in MSC, II (iii), 1931.

*The Comedy of the Prodigal Son*, retranslation by R. Simpson from the German version of a lost English play (printed in German 1620), in R. Simpson, *The School of Shakespeare*, II (1878). See J. Tittmann, *Die Schauspiele der englischen Komödianten* (Leipzig, 1880).

*Rare Triumphs of Love and Fortune.* ? Acted at court 1582; printed 1589, ed. J. P. Collier (*Five Old Plays*, 1851), H-D 6, MSR, 1931.

*A merye enterlude entitled Respublica* (written 1553). MS. first printed by J. P. Collier, *Illustrations of English Literature*, I (1866); Brandl; ed. L. A. Magnus, EETS E.S. 94 (1905)—superseded by ed. W. W. Greg, EETS O.S. 226 (1952); ed. Farmer, 1907, 1908*. Greg (op. cit.), Leicester Bradner, *MLN*, 1927, and T. W. Baldwin, *Five-Act Structure*, are among those who argue for N. Udall's authorship of this play.

*A newe playe [of Robyn Hoode] for to be played in Maye games*, appended to *A mery geste of Robyn Hoode*, *c.* 1560. Ed. Farmer, 1914, MSC I (ii), 1909.

*Somebody and Others* or *The Spoiling of Lady Verity.* (Written *c.* 1550?; published 1550?) Two leaves survive of this translation of French Protestant morality *La Vérité Cachée*. See S. R. Maitland, *List of Early Printed Books at Lambeth* (1843); MSC, II (iii), 1931.

*Temperance and Humility*. One leaf survives (written 1535?; published 1537?). Ed. MSC, I (iii), 1909. See T. W. Craik in *RES*, 1953.

*Thersites* (written 1537; published n.d. [1562?]). Ed. F. J. Child (*Four Old Plays*) 1848; E. W. Ashbee, 1876?* H-D 1, Farmer, 1905, 1912*; Pollard, *English Miracle Plays* (abridged). A. W. Pollard (Gayley, pp. 12–16) tries to attribute the play to Heywood. A. R. Moon, *The Library*, 1926, tries, on inadequate evidence, to give it to N. Udall. Holthausen prints the original Textor dialogue in *E. Studien*, 1902. Boas (*University Drama*) points to its Oxford references; and Pollard (*TLS*, 1918, p. 337) and Boas (*TLS*, 1918, p. 349) renew this information.

*Tom Tyler and his Wife*. First surviving text is called *An Excellent Old Play, as it was printed and acted about a hundred years ago. The Second impression* . . . 1661. Written *c.* 1560?; there is an S.R. entry in 1563 of a 'ballett of tom Tyler' which may refer to this play. Ed. F. E. Schelling, *PMLA*, 1900; Farmer, 1906; MSR, 1910; Farmer, 1912*.

*Trial of Treasure*. Written 1565?; published 1567. Ed. Halliwell (Percy Society, 1842); H-D 3; Farmer, 1906, 1908*; L. M. Oliver has argued that the play is by William Wager (*HLQ*, 1945/46), on the basis of parallels first pointed out by L. B. Wright, *Anglia*, 1930 (p. 116, n. 1). E. B. Daw (*MP*, 1917/18) points to source-material in Foxe's *Acts and Monuments*.

*The Wars of Cyrus*. Written either 1576–80, or else after 1587; published 1594. Ed. W. Keller, *Sh. Jb.*, 1901; Farmer, 1911*; ed. J. P. Brawner (1942). W. J. Lawrence (*TLS*, 11 August 1921) first pointed to Richard Farrant as possible author, on the basis of a manuscript song first described by G. E. P. Arkwright (*NQ*, 5 May 1906) and again in 'Elizabethan Choirboys and their music', *Proc. Musical Assoc.* (1913/14). Fellowes (*Grove* (1954)) picks up his point that in a BM. manuscript this song is assigned to Robert Parsons, but unlike Arkwright seems inclined to believe this. Brawner's edition develops the thesis of

Farrant's authorship; it is accepted by F. S. Boas—*TLS*, 31 April 1945—and Irving Ribner, '*Tamburlaine* and *The Wars of Cyrus*', *JEGP*, 1954. G. K. Hunter (*NQ*, 1961, pp. 395 et seq.) suggests a date later than Farrant.

*Wealth and Health*. Written 1554?; S.R. August/September 1557; published n.d. [1557?]. Ed. Farmer, 1907, 1907*; MSR, 1907 (corrected in MSC, I (i) (1908)); ed. F. Holthausen, 1908 (revised edn. 1922). Sir Mark Hunter (*MLR*, 1907/8) argues for a date of composition at the end of the fifteenth century; A. E. H. Swaen, *E. Studien*, 1909/10, explicates Dutch passages; T. W. Craik, *RES*, 1953, argues that the play has a Marian origin.

*The World and the Child* (*Mundus et Infans*). (Written ?; published 1522). Ed. anon. (Roxburghe Club) 1817, Dodsley, H-D 1, Manly I, Farmer, 1905, 1909*, ed. John Hampden, 1935. H. N. MacCracken, *PMLA*, 1908, points to a close analogue. See Bevington, pp. 116 ff., etc.

*Youth*. Written *c*. 1520?; entered S.R. Aug.–Sept. 1557. Published (1) n.d. [*c*. 1530?]—fragment; (2) n.d. [1557]; (3) n.d. [1562?]. Ed. Halliwell 1849; H-D 2, Materialien, 1905, Farmer, 1906, 1909* (first two editions). See S. R. Maitland, *A List of Some of the Early Printed Books at Lambeth* (1843). Creizenach (*Geschichte*) and McKerrow (Materialien, 1905) discuss the relationship with *Hycke Scorner*, as does E. T. Schell, *PQ*, 1966.

# INDEX